MULTIPLICATION FAST FACTS

Timed Math Facts Practice Tests

This book belongs to:

Welcome to *Multiplication Fast Facts!* This workbook is designed to help you master multiplication by covering all facts with the factors from 0 to 12 using 100 timed tests. By practicing regularly, you'll strengthen your multiplication skills, increase your speed, and build the confidence you need to succeed in math.

How to Use This Workbook:

- Set a Timer: Each test is designed to be completed as quickly as possible. Use a timer to track your speed and challenge yourself to improve.
- Practice Regularly: Consistency is key! Work on one or two tests each day to steadily improve your skills.
- Review Mistakes: After each test, check the answer key in the back, review any errors, and focus on the facts that need extra practice.
- Track Your Progress: At the top of each page is space to record your time and score. In the back of the book, you will also find a progress tracking chart you can use to record your growth over time. Use it to see how far you've come and to set goals for improvement. Remember, progress takes time—keep practicing, and you'll see great results!

Good luck, and have fun mastering your multiplication facts!

Printed in the United States of America
ISBN: 978-1-64824-045-4
EDventure Learning LLC
5601 State Route 31 #1296
Clay, NY 13039

www.edventurelearning.com

Email us at hello@edventurelearning.com
Follow us on social media @edventurelearning

CONTENTS

1

Multiply by 0 or 1

Time: ________

Score: _____/ 40

(1) 0×3	(2) 7×0	(3) 4×0	(4) 1×0	(5) 1×9
(6) 8×0	(7) 0×7	(8) 1×1	(9) 7×1	(10) 6×0
(11) 0×4	(12) 0×1	(13) 0×0	(14) 1×5	(15) 1×6
(16) 5×0	(17) 3×1	(18) 1×8	(19) 0×2	(20) 9×0
(21) 5×1	(22) 0×6	(23) 0×5	(24) 3×0	(25) 9×1
(26) 7×0	(27) 6×1	(28) 4×0	(29) 0×0	(30) 8×1
(31) 0×7	(32) 1×7	(33) 2×0	(34) 1×0	(35) 1×3
(36) 0×9	(37) 0×8	(38) 0×1	(39) 0×4	(40) 1×1

2

Multiply by 0 or 1

Time: ________

Score: _____/ 40

1. 4×0	2. 3×1	3. 0×1	4. 0×6	5. 5×1
6. 7×1	7. 1×4	8. 0×2	9. 8×0	10. 1×6
11. 0×0	12. 0×4	13. 1×1	14. 4×1	15. 0×9
16. 5×0	17. 3×0	18. 9×1	19. 1×8	20. 1×0
21. 2×0	22. 1×5	23. 1×7	24. 1×1	25. 1×9
26. 0×5	27. 6×1	28. 4×0	29. 0×1	30. 1×4
31. 9×0	32. 0×4	33. 0×0	34. 4×1	35. 8×1
36. 0×3	37. 6×0	38. 1×3	39. 1×0	40. 0×8

3

Multiply by 0 or 1

Time: __________

Score: _____/ 40

(1) 1×6	(2) 2×0	(3) 8×1	(4) 5×1	(5) 7×1
(6) 1×0	(7) 0×3	(8) 9×0	(9) 5×0	(10) 0×0
(11) 9×1	(12) 1×4	(13) 6×0	(14) 0×1	(15) 0×4
(16) 4×1	(17) 3×1	(18) 1×1	(19) 4×0	(20) 8×0
(21) 1×0	(22) 0×1	(23) 1×4	(24) 0×5	(25) 1×3
(26) 3×0	(27) 1×1	(28) 0×4	(29) 0×9	(30) 0×8
(31) 1×7	(32) 0×6	(33) 0×0	(34) 1×8	(35) 1×9
(36) 4×1	(37) 6×1	(38) 1×5	(39) 0×2	(40) 4×0

4

Multiply by 0 or 1

Time: ___________

Score: _____/ 40

(1) 1×1	(2) 1×5	(3) 3×1	(4) 2×1	(5) 1×7
(6) 4×1	(7) 1×0	(8) 8×0	(9) 0×2	(10) 6×0
(11) 6×1	(12) 5×0	(13) 0×1	(14) 1×4	(15) 1×2
(16) 0×3	(17) 1×8	(18) 9×1	(19) 0×0	(20) 9×0
(21) 4×1	(22) 1×3	(23) 3×0	(24) 2×1	(25) 0×8
(26) 5×1	(27) 8×1	(28) 0×5	(29) 7×1	(30) 1×4
(31) 2×0	(32) 1×9	(33) 0×9	(34) 1×2	(35) 1×0
(36) 0×0	(37) 0×6	(38) 0×1	(39) 1×1	(40) 1×6

5

Multiply by 2

Time: ________

Score: _____/ 40

(1) 0 × 2	(2) 4 × 2	(3) 5 × 2	(4) 1 × 2	(5) 9 × 2
(6) 2 × 8	(7) 3 × 2	(8) 2 × 7	(9) 2 × 2	(10) 2 × 6
(11) 2 × 5	(12) 7 × 2	(13) 2 × 3	(14) 6 × 2	(15) 2 × 4
(16) 2 × 0	(17) 2 × 2	(18) 2 × 9	(19) 8 × 2	(20) 2 × 1
(21) 2 × 6	(22) 4 × 2	(23) 9 × 2	(24) 2 × 8	(25) 0 × 2
(26) 1 × 2	(27) 2 × 2	(28) 3 × 2	(29) 2 × 7	(30) 5 × 2
(31) 8 × 2	(32) 6 × 2	(33) 2 × 4	(34) 2 × 0	(35) 7 × 2
(36) 2 × 9	(37) 2 × 2	(38) 2 × 5	(39) 2 × 3	(40) 2 × 1

6

Multiply by 2

Time: ________

Score: _____/ 40

(1) 4×2	(2) 2×2	(3) 2×9	(4) 2×0	(5) 7×2
(6) 2×6	(7) 2×8	(8) 2×3	(9) 5×2	(10) 1×2
(11) 6×2	(12) 9×2	(13) 3×2	(14) 2×5	(15) 8×2
(16) 0×2	(17) 2×2	(18) 2×7	(19) 2×1	(20) 2×4
(21) 2×3	(22) 5×2	(23) 1×2	(24) 4×2	(25) 2×8
(26) 7×2	(27) 2×6	(28) 2×9	(29) 2×0	(30) 2×2
(31) 0×2	(32) 2×1	(33) 2×7	(34) 2×5	(35) 9×2
(36) 6×2	(37) 8×2	(38) 2×2	(39) 3×2	(40) 2×4

7

Multiply by 2

Time: ________

Score: _____/ 40

(1) 1×2	(2) 8×2	(3) 2×6	(4) 2×4	(5) 2×2
(6) 7×2	(7) 2×9	(8) 0×2	(9) 2×5	(10) 2×3
(11) 2×7	(12) 4×2	(13) 6×2	(14) 2×8	(15) 3×2
(16) 2×1	(17) 5×2	(18) 2×0	(19) 9×2	(20) 2×2
(21) 2×3	(22) 0×2	(23) 1×2	(24) 2×4	(25) 2×6
(26) 2×5	(27) 2×2	(28) 8×2	(29) 7×2	(30) 2×9
(31) 2×7	(32) 3×2	(33) 2×8	(34) 2×1	(35) 6×2
(36) 2×2	(37) 4×2	(38) 9×2	(39) 5×2	(40) 2×0

8

Multiply by 2

Time: __________

Score: _____/ 40

(1) 2 × 9	(2) 3 × 2	(3) 0 × 2	(4) 2 × 6	(5) 1 × 2
(6) 7 × 2	(7) 2 × 4	(8) 2 × 5	(9) 2 × 8	(10) 2 × 2
(11) 9 × 2	(12) 2 × 3	(13) 2 × 0	(14) 2 × 2	(15) 5 × 2
(16) 2 × 7	(17) 6 × 2	(18) 8 × 2	(19) 4 × 2	(20) 2 × 1
(21) 2 × 8	(22) 2 × 2	(23) 2 × 6	(24) 2 × 4	(25) 3 × 2
(26) 0 × 2	(27) 2 × 5	(28) 1 × 2	(29) 2 × 9	(30) 7 × 2
(31) 9 × 2	(32) 4 × 2	(33) 2 × 3	(34) 2 × 0	(35) 8 × 2
(36) 6 × 2	(37) 5 × 2	(38) 2 × 7	(39) 2 × 1	(40) 2 × 2

9

Multiply by 3

Time: ____________

Score: ______/ 40

(1) 6×3	(2) 3×4	(3) 5×3	(4) 3×2	(5) 3×0
(6) 7×3	(7) 3×1	(8) 3×3	(9) 8×3	(10) 3×9
(11) 3×7	(12) 4×3	(13) 1×3	(14) 3×5	(15) 9×3
(16) 0×3	(17) 3×6	(18) 2×3	(19) 3×8	(20) 3×3
(21) 7×3	(22) 3×1	(23) 3×0	(24) 3×2	(25) 6×3
(26) 5×3	(27) 3×3	(28) 3×9	(29) 8×3	(30) 3×4
(31) 9×3	(32) 3×6	(33) 3×8	(34) 4×3	(35) 1×3
(36) 3×3	(37) 3×7	(38) 2×3	(39) 3×5	(40) 0×3

10

Multiply by 3

Time: ______

Score: _____/ 40

(1) 3 × 4	(2) 3 × 3	(3) 3 × 7	(4) 2 × 3	(5) 3 × 1
(6) 9 × 3	(7) 8 × 3	(8) 5 × 3	(9) 6 × 3	(10) 3 × 0
(11) 3 × 6	(12) 3 × 9	(13) 3 × 5	(14) 0 × 3	(15) 7 × 3
(16) 4 × 3	(17) 3 × 2	(18) 3 × 8	(19) 1 × 3	(20) 3 × 3
(21) 2 × 3	(22) 3 × 3	(23) 8 × 3	(24) 3 × 4	(25) 5 × 3
(26) 3 × 0	(27) 6 × 3	(28) 9 × 3	(29) 3 × 1	(30) 3 × 7
(31) 4 × 3	(32) 7 × 3	(33) 3 × 5	(34) 3 × 9	(35) 0 × 3
(36) 3 × 6	(37) 3 × 8	(38) 1 × 3	(39) 3 × 3	(40) 3 × 2

11

Multiply by 3

Time: ________

Score: _____/ 40

(1) 3×0	(2) 3×5	(3) 3×3	(4) 3×8	(5) 7×3
(6) 9×3	(7) 3×4	(8) 3×2	(9) 3×6	(10) 3×1
(11) 6×3	(12) 4×3	(13) 5×3	(14) 2×3	(15) 1×3
(16) 3×9	(17) 0×3	(18) 3×7	(19) 3×3	(20) 8×3
(21) 3×6	(22) 3×2	(23) 9×3	(24) 3×1	(25) 3×4
(26) 3×0	(27) 3×8	(28) 3×5	(29) 3×3	(30) 7×3
(31) 8×3	(32) 0×3	(33) 1×3	(34) 4×3	(35) 3×7
(36) 2×3	(37) 6×3	(38) 3×3	(39) 5×3	(40) 3×9

12

Multiply by 3

Time: __________

Score: _____/ 40

(1) 7 × 3	(2) 3 × 9	(3) 3 × 3	(4) 8 × 3	(5) 3 × 4
(6) 5 × 3	(7) 3 × 1	(8) 2 × 3	(9) 0 × 3	(10) 3 × 6
(11) 3 × 5	(12) 3 × 7	(13) 6 × 3	(14) 3 × 2	(15) 3 × 8
(16) 1 × 3	(17) 9 × 3	(18) 3 × 3	(19) 3 × 0	(20) 4 × 3
(21) 7 × 3	(22) 3 × 4	(23) 3 × 9	(24) 8 × 3	(25) 0 × 3
(26) 3 × 1	(27) 5 × 3	(28) 3 × 6	(29) 2 × 3	(30) 3 × 3
(31) 3 × 7	(32) 6 × 3	(33) 3 × 0	(34) 3 × 2	(35) 1 × 3
(36) 3 × 5	(37) 3 × 8	(38) 4 × 3	(39) 3 × 3	(40) 9 × 3

13

Multiply by 0-3

Time: ____________

Score: _____/ 40

(1) 3×1	(2) 1×2	(3) 9×0	(4) 6×0	(5) 0×3
(6) 3×3	(7) 1×1	(8) 9×3	(9) 0×1	(10) 0×6
(11) 4×1	(12) 2×5	(13) 4×3	(14) 5×0	(15) 2×3
(16) 6×3	(17) 7×1	(18) 4×2	(19) 7×2	(20) 3×7
(21) 3×2	(22) 3×4	(23) 1×0	(24) 6×2	(25) 3×8
(26) 1×3	(27) 2×1	(28) 2×4	(29) 2×0	(30) 3×6
(31) 3×5	(32) 2×9	(33) 0×2	(34) 0×0	(35) 1×5
(36) 1×2	(37) 1×4	(38) 3×0	(39) 2×3	(40) 1×9

14

Multiply by 0-3

Time: __________

Score: _____/ 40

(1) 2×0	(2) 0×3	(3) 3×7	(4) 1×4	(5) 0×4
(6) 9×1	(7) 3×4	(8) 0×2	(9) 1×3	(10) 4×0
(11) 2×2	(12) 2×7	(13) 4×3	(14) 3×0	(15) 3×9
(16) 9×0	(17) 3×8	(18) 1×5	(19) 1×0	(20) 3×6
(21) 1×1	(22) 0×0	(23) 3×2	(24) 0×4	(25) 3×3
(26) 1×7	(27) 3×1	(28) 6×2	(29) 2×3	(30) 2×2
(31) 0×1	(32) 6×3	(33) 4×2	(34) 2×5	(35) 3×2
(36) 4×1	(37) 3×5	(38) 2×9	(39) 2×4	(40) 0×5

15

Multiply by 0-3

Time: ________

Score: _____/ 40

(1) 1×9	(2) 1×6	(3) 5×2	(4) 4×1	(5) 0×3
(6) 7×3	(7) 7×1	(8) 2×3	(9) 0×0	(10) 0×6
(11) 6×3	(12) 2×7	(13) 8×1	(14) 3×1	(15) 4×0
(16) 1×4	(17) 5×3	(18) 1×0	(19) 6×1	(20) 2×0
(21) 0×2	(22) 3×2	(23) 1×5	(24) 6×0	(25) 2×3
(26) 3×6	(27) 0×1	(28) 0×4	(29) 1×1	(30) 3×8
(31) 1×8	(32) 0×9	(33) 2×9	(34) 6×2	(35) 3×0
(36) 1×3	(37) 3×3	(38) 3×9	(39) 0×5	(40) 4×0

16

Multiply by 0-3

Time: __________

Score: _____/ 40

(1) 2×2	(2) 7×1	(3) 0×3	(4) 5×1	(5) 0×0
(6) 7×3	(7) 3×1	(8) 3×3	(9) 4×1	(10) 2×1
(11) 0×5	(12) 4×3	(13) 6×1	(14) 0×8	(15) 8×0
(16) 0×2	(17) 6×2	(18) 2×8	(19) 3×4	(20) 2×0
(21) 1×2	(22) 8×3	(23) 0×1	(24) 1×6	(25) 0×8
(26) 0×9	(27) 1×1	(28) 5×3	(29) 1×4	(30) 1×3
(31) 9×2	(32) 3×0	(33) 2×2	(34) 8×2	(35) 1×9
(36) 2×5	(37) 1×0	(38) 9×3	(39) 7×2	(40) 2×1

17

Multiply by 0-3

Time: ________

Score: _____/ 40

(1) 8×0	(2) 4×1	(3) 6×0	(4) 1×3	(5) 0×5
(6) 0×2	(7) 3×5	(8) 7×1	(9) 0×6	(10) 3×4
(11) 3×1	(12) 8×3	(13) 4×3	(14) 3×6	(15) 0×9
(16) 3×3	(17) 2×0	(18) 1×4	(19) 2×1	(20) 3×7
(21) 5×2	(22) 1×6	(23) 2×6	(24) 6×3	(25) 0×8
(26) 9×1	(27) 9×2	(28) 1×0	(29) 8×0	(30) 6×1
(31) 1×2	(32) 0×0	(33) 0×3	(34) 5×1	(35) 1×1
(36) 7×2	(37) 3×0	(38) 0×1	(39) 2×1	(40) 3×9

18

Multiply by 0-3

Time: ____________

Score: _____/ 40

(1) 9×1	(2) 2×8	(3) 0×2	(4) 7×3	(5) 3×8
(6) 2×5	(7) 1×0	(8) 3×0	(9) 1×8	(10) 8×2
(11) 2×0	(12) 1×1	(13) 0×0	(14) 2×2	(15) 0×4
(16) 2×9	(17) 8×1	(18) 9×3	(19) 0×1	(20) 3×5
(21) 0×7	(22) 3×1	(23) 2×6	(24) 7×0	(25) 4×0
(26) 2×3	(27) 5×1	(28) 0×3	(29) 2×2	(30) 0×4
(31) 3×3	(32) 1×7	(33) 2×7	(34) 0×9	(35) 4×2
(36) 5×0	(37) 3×2	(38) 1×3	(39) 2×3	(40) 2×4

19

Multiply by 0-3

Time: ________

Score: _____/ 40

① 1×0	② 1×7	③ 4×2	④ 3×5	⑤ 7×0
⑥ 2×2	⑦ 0×3	⑧ 7×3	⑨ 3×0	⑩ 5×2
⑪ 0×0	⑫ 2×2	⑬ 4×3	⑭ 1×1	⑮ 1×9
⑯ 0×7	⑰ 0×4	⑱ 2×0	⑲ 0×2	⑳ 2×7
(21) 1×5	(22) 2×3	(23) 4×0	(24) 2×4	(25) 2×9
(26) 9×3	(27) 0×1	(28) 3×3	(29) 0×9	(30) 3×4
(31) 3×2	(32) 2×6	(33) 0×5	(34) 3×6	(35) 3×1
(36) 1×3	(37) 0×4	(38) 6×3	(39) 8×3	(40) 2×3

20

Multiply by 0-3

Time: ________

Score: _____/ 40

(1) 1×2	(2) 3×0	(3) 2×0	(4) 6×1	(5) 3×9
(6) 6×0	(7) 9×2	(8) 5×2	(9) 2×1	(10) 0×0
(11) 0×8	(12) 3×8	(13) 3×3	(14) 7×3	(15) 0×7
(16) 2×6	(17) 3×4	(18) 8×0	(19) 5×1	(20) 9×0
(21) 8×2	(22) 1×9	(23) 7×1	(24) 3×5	(25) 3×1
(26) 0×1	(27) 1×2	(28) 7×0	(29) 0×6	(30) 0×8
(31) 0×3	(32) 1×0	(33) 4×3	(34) 1×1	(35) 0×2
(36) 1×3	(37) 2×8	(38) 5×0	(39) 1×6	(40) 2×7

21

Multiply by 4

Time: ____________

Score: _____/ 40

(1) 4 × 7	(2) 5 × 4	(3) 4 × 4	(4) 3 × 4	(5) 8 × 4
(6) 4 × 1	(7) 4 × 0	(8) 4 × 2	(9) 4 × 9	(10) 6 × 4
(11) 4 × 3	(12) 4 × 5	(13) 0 × 4	(14) 1 × 4	(15) 4 × 8
(16) 4 × 4	(17) 2 × 4	(18) 4 × 6	(19) 7 × 4	(20) 9 × 4
(21) 3 × 4	(22) 4 × 9	(23) 4 × 2	(24) 8 × 4	(25) 4 × 1
(26) 4 × 4	(27) 4 × 7	(28) 4 × 0	(29) 5 × 4	(30) 6 × 4
(31) 9 × 4	(32) 1 × 4	(33) 4 × 8	(34) 4 × 5	(35) 4 × 3
(36) 7 × 4	(37) 2 × 4	(38) 0 × 4	(39) 4 × 6	(40) 4 × 4

22

Multiply by 4

Time: ________

Score: _____/ 40

(1) 4 × 8	(2) 4 × 4	(3) 4 × 9	(4) 5 × 4	(5) 3 × 4
(6) 7 × 4	(7) 4 × 6	(8) 1 × 4	(9) 2 × 4	(10) 4 × 0
(11) 4 × 2	(12) 0 × 4	(13) 4 × 7	(14) 4 × 3	(15) 6 × 4
(16) 9 × 4	(17) 8 × 4	(18) 4 × 5	(19) 4 × 4	(20) 4 × 1
(21) 2 × 4	(22) 4 × 0	(23) 4 × 6	(24) 4 × 4	(25) 3 × 4
(26) 7 × 4	(27) 4 × 9	(28) 4 × 8	(29) 1 × 4	(30) 5 × 4
(31) 4 × 7	(32) 4 × 5	(33) 4 × 3	(34) 9 × 4	(35) 4 × 2
(36) 0 × 4	(37) 4 × 1	(38) 4 × 4	(39) 6 × 4	(40) 8 × 4

23

Multiply by 4

Time: ________

Score: _____/ 40

(1) 6×4	(2) 4×8	(3) 3×4	(4) 1×4	(5) 4×2
(6) 4×0	(7) 5×4	(8) 4×9	(9) 4×4	(10) 4×7
(11) 0×4	(12) 2×4	(13) 8×4	(14) 4×4	(15) 4×6
(16) 4×1	(17) 7×4	(18) 9×4	(19) 4×5	(20) 4×3
(21) 6×4	(22) 3×4	(23) 4×7	(24) 4×8	(25) 4×2
(26) 1×4	(27) 4×9	(28) 5×4	(29) 4×0	(30) 4×4
(31) 4×6	(32) 4×4	(33) 0×4	(34) 2×4	(35) 4×1
(36) 4×3	(37) 9×4	(38) 7×4	(39) 4×5	(40) 8×4

24

Multiply by 4

Time: __________

Score: _____/ 40

① 1×4	② 6×4	③ 4×7	④ 4×4	⑤ 8×4
⑥ 4×3	⑦ 4×9	⑧ 2×4	⑨ 5×4	⑩ 0×4
⑪ 9×4	⑫ 7×4	⑬ 4×8	⑭ 4×2	⑮ 4×1
⑯ 4×0	⑰ 4×6	⑱ 3×4	⑲ 4×4	⑳ 4×5
21 8×4	22 0×4	23 5×4	24 4×4	25 4×7
26 1×4	27 2×4	28 6×4	29 4×9	30 4×3
31 4×1	32 3×4	33 4×5	34 4×6	35 9×4
36 4×0	37 7×4	38 4×4	39 4×2	40 4×8

25

Multiply by 5

Time: ________

Score: _____/ 40

(1) 5 × 9	(2) 5 × 5	(3) 5 × 1	(4) 5 × 0	(5) 5 × 2
(6) 7 × 5	(7) 6 × 5	(8) 5 × 3	(9) 5 × 4	(10) 5 × 8
(11) 9 × 5	(12) 0 × 5	(13) 8 × 5	(14) 5 × 7	(15) 5 × 5
(16) 5 × 6	(17) 1 × 5	(18) 2 × 5	(19) 3 × 5	(20) 4 × 5
(21) 5 × 0	(22) 5 × 8	(23) 5 × 2	(24) 5 × 3	(25) 6 × 5
(26) 5 × 1	(27) 5 × 4	(28) 5 × 5	(29) 7 × 5	(30) 5 × 9
(31) 8 × 5	(32) 5 × 6	(33) 9 × 5	(34) 2 × 5	(35) 1 × 5
(36) 4 × 5	(37) 5 × 5	(38) 5 × 7	(39) 3 × 5	(40) 0 × 5

26

Multiply by 5

Time: ________

Score: _____/ 40

(1) 5×7	(2) 5×0	(3) 5×1	(4) 8×5	(5) 5×9
(6) 5×3	(7) 2×5	(8) 5×5	(9) 5×6	(10) 5×4
(11) 9×5	(12) 1×5	(13) 5×5	(14) 7×5	(15) 5×8
(16) 6×5	(17) 4×5	(18) 3×5	(19) 5×2	(20) 0×5
(21) 5×7	(22) 5×6	(23) 5×0	(24) 5×3	(25) 2×5
(26) 5×4	(27) 5×9	(28) 8×5	(29) 5×1	(30) 5×5
(31) 7×5	(32) 5×5	(33) 5×2	(34) 5×8	(35) 6×5
(36) 1×5	(37) 3×5	(38) 4×5	(39) 0×5	(40) 9×5

27

Multiply by 5

Time: ________

Score: _____/ 40

(1) 1 × 5	(2) 6 × 5	(3) 5 × 7	(4) 5 × 5	(5) 5 × 4
(6) 5 × 3	(7) 0 × 5	(8) 5 × 8	(9) 9 × 5	(10) 2 × 5
(11) 5 × 5	(12) 5 × 0	(13) 5 × 2	(14) 5 × 9	(15) 7 × 5
(16) 8 × 5	(17) 4 × 5	(18) 5 × 6	(19) 3 × 5	(20) 5 × 1
(21) 5 × 4	(22) 5 × 7	(23) 2 × 5	(24) 9 × 5	(25) 6 × 5
(26) 1 × 5	(27) 5 × 3	(28) 5 × 8	(29) 0 × 5	(30) 5 × 5
(31) 5 × 9	(32) 8 × 5	(33) 5 × 2	(34) 5 × 5	(35) 4 × 5
(36) 7 × 5	(37) 5 × 1	(38) 3 × 5	(39) 5 × 0	(40) 5 × 6

28

Multiply by 5

Time: __________

Score: _____/ 40

(1) 0×5	(2) 5×8	(3) 9×5	(4) 7×5	(5) 5×1
(6) 5×5	(7) 4×5	(8) 5×2	(9) 6×5	(10) 5×3
(11) 5×0	(12) 3×5	(13) 5×5	(14) 2×5	(15) 5×7
(16) 5×9	(17) 5×4	(18) 8×5	(19) 1×5	(20) 5×6
(21) 4×5	(22) 7×5	(23) 5×1	(24) 0×5	(25) 9×5
(26) 5×2	(27) 5×5	(28) 5×8	(29) 6×5	(30) 5×3
(31) 5×0	(32) 3×5	(33) 2×5	(34) 5×5	(35) 5×9
(36) 1×5	(37) 5×4	(38) 5×7	(39) 8×5	(40) 5×6

29

Multiply by 6

Time: __________

Score: _____/ 40

1. 2 × 6	2. 6 × 9	3. 1 × 6	4. 0 × 6	5. 6 × 4
6. 5 × 6	7. 6 × 7	8. 6 × 6	9. 3 × 6	10. 8 × 6
11. 6 × 1	12. 4 × 6	13. 6 × 2	14. 6 × 5	15. 6 × 8
16. 6 × 6	17. 9 × 6	18. 7 × 6	19. 6 × 3	20. 6 × 0
21. 6 × 6	22. 5 × 6	23. 1 × 6	24. 6 × 9	25. 0 × 6
26. 8 × 6	27. 2 × 6	28. 6 × 7	29. 3 × 6	30. 6 × 4
31. 6 × 5	32. 4 × 6	33. 6 × 8	34. 6 × 1	35. 9 × 6
36. 6 × 2	37. 7 × 6	38. 6 × 0	39. 6 × 3	40. 6 × 6

30

Multiply by 6

Time: ________

Score: _____/ 40

(1) 6×5	(2) 6×7	(3) 0×6	(4) 6×3	(5) 6×1
(6) 6×9	(7) 6×6	(8) 4×6	(9) 6×2	(10) 6×8
(11) 1×6	(12) 7×6	(13) 6×4	(14) 5×6	(15) 6×6
(16) 2×6	(17) 6×0	(18) 3×6	(19) 9×6	(20) 8×6
(21) 6×2	(22) 4×6	(23) 6×5	(24) 6×8	(25) 6×9
(26) 6×7	(27) 6×6	(28) 6×1	(29) 6×3	(30) 0×6
(31) 7×6	(32) 9×6	(33) 2×6	(34) 6×4	(35) 1×6
(36) 5×6	(37) 3×6	(38) 6×0	(39) 8×6	(40) 6×6

31

Multiply by 6

Time: ________

Score: _____/ 40

(1) 5×6	(2) 6×6	(3) 6×0	(4) 6×1	(5) 6×9
(6) 4×6	(7) 2×6	(8) 6×7	(9) 8×6	(10) 6×3
(11) 6×8	(12) 3×6	(13) 9×6	(14) 6×5	(15) 6×2
(16) 6×6	(17) 6×4	(18) 7×6	(19) 1×6	(20) 0×6
(21) 6×7	(22) 6×3	(23) 6×0	(24) 6×9	(25) 8×6
(26) 5×6	(27) 2×6	(28) 6×1	(29) 4×6	(30) 6×6
(31) 7×6	(32) 1×6	(33) 6×4	(34) 6×6	(35) 6×8
(36) 6×2	(37) 9×6	(38) 6×5	(39) 3×6	(40) 0×6

32

Multiply by 6

Time: ________

Score: _____/ 40

1. 6 × 2	2. 9 × 6	3. 6 × 5	4. 0 × 6	5. 4 × 6
6. 8 × 6	7. 1 × 6	8. 6 × 7	9. 6 × 6	10. 3 × 6
11. 7 × 6	12. 2 × 6	13. 6 × 1	14. 5 × 6	15. 6 × 4
16. 6 × 3	17. 6 × 6	18. 6 × 8	19. 6 × 9	20. 6 × 0
21. 4 × 6	22. 3 × 6	23. 6 × 6	24. 6 × 2	25. 6 × 5
26. 6 × 7	27. 8 × 6	28. 9 × 6	29. 0 × 6	30. 1 × 6
31. 2 × 6	32. 6 × 9	33. 6 × 0	34. 5 × 6	35. 6 × 4
36. 6 × 6	37. 6 × 1	38. 7 × 6	39. 6 × 3	40. 6 × 8

33

Multiply by 4-6

Time: ________

Score: _____/ 40

(1) 0 × 6	(2) 5 × 2	(3) 3 × 4	(4) 4 × 4	(5) 1 × 4
(6) 7 × 6	(7) 8 × 6	(8) 5 × 8	(9) 6 × 5	(10) 6 × 4
(11) 4 × 9	(12) 4 × 8	(13) 4 × 5	(14) 4 × 6	(15) 7 × 4
(16) 7 × 5	(17) 6 × 3	(18) 5 × 6	(19) 5 × 4	(20) 3 × 5
(21) 2 × 4	(22) 5 × 9	(23) 6 × 2	(24) 0 × 4	(25) 0 × 5
(26) 1 × 5	(27) 5 × 5	(28) 9 × 6	(29) 6 × 6	(30) 4 × 2
(31) 6 × 5	(32) 5 × 7	(33) 5 × 0	(34) 4 × 5	(35) 8 × 5
(36) 2 × 4	(37) 5 × 5	(38) 5 × 6	(39) 8 × 4	(40) 6 × 7

34

Multiply by 4-6

Time: ________

Score: _____/ 40

① 2×5	② 6×6	③ 7×4	④ 5×1	⑤ 5×3
⑥ 6×3	⑦ 5×4	⑧ 0×4	⑨ 0×6	⑩ 9×4
⑪ 6×9	⑫ 2×6	⑬ 4×1	⑭ 2×4	⑮ 6×4
⑯ 3×4	⑰ 4×4	⑱ 6×8	⑲ 4×6	⑳ 5×9
(21) 8×6	(22) 4×4	(23) 5×2	(24) 4×9	(25) 0×5
(26) 6×4	(27) 4×5	(28) 8×5	(29) 4×7	(30) 4×3
(31) 9×5	(32) 5×4	(33) 1×4	(34) 6×5	(35) 5×7
(36) 6×2	(37) 6×7	(38) 3×5	(39) 5×6	(40) 4×0

35

Multiply by 4-6

Time: ________

Score: _____/ 40

(1) 1×5	(2) 4×6	(3) 6×9	(4) 1×6	(5) 6×1
(6) 6×0	(7) 4×8	(8) 5×5	(9) 6×6	(10) 3×6
(11) 8×4	(12) 0×6	(13) 6×4	(14) 5×6	(15) 5×3
(16) 4×4	(17) 6×5	(18) 7×4	(19) 5×2	(20) 6×8
(21) 6×6	(22) 1×6	(23) 5×5	(24) 9×6	(25) 0×4
(26) 9×5	(27) 4×3	(28) 5×1	(29) 4×6	(30) 5×8
(31) 0×5	(32) 6×1	(33) 9×4	(34) 5×7	(35) 5×4
(36) 7×6	(37) 6×3	(38) 1×4	(39) 4×5	(40) 2×6

36

Multiply by 4-6

Time: ________

Score: _____/ 40

(1) 6×4	(2) 8×5	(3) 6×3	(4) 9×6	(5) 4×8
(6) 5×1	(7) 5×0	(8) 9×4	(9) 6×0	(10) 8×6
(11) 6×7	(12) 6×5	(13) 5×5	(14) 9×5	(15) 4×5
(16) 4×6	(17) 7×4	(18) 5×7	(19) 5×3	(20) 0×4
(21) 5×2	(22) 5×4	(23) 6×2	(24) 2×4	(25) 4×4
(26) 1×4	(27) 5×6	(28) 4×2	(29) 4×3	(30) 6×6
(31) 6×9	(32) 4×1	(33) 6×4	(34) 6×5	(35) 3×6
(36) 0×5	(37) 4×7	(38) 4×6	(39) 5×6	(40) 7×5

37

Multiply by 4-6

Time: ___________

Score: _____/ 40

(1) 4×9	(2) 5×3	(3) 4×3	(4) 5×4	(5) 4×5
(6) 6×6	(7) 4×2	(8) 2×5	(9) 0×4	(10) 4×4
(11) 6×0	(12) 5×9	(13) 6×2	(14) 2×4	(15) 5×1
(16) 8×6	(17) 8×5	(18) 7×6	(19) 4×8	(20) 5×5
(21) 6×5	(22) 8×4	(23) 9×4	(24) 6×7	(25) 6×6
(26) 3×4	(27) 4×0	(28) 5×4	(29) 7×5	(30) 4×1
(31) 5×5	(32) 4×2	(33) 9×6	(34) 6×4	(35) 6×8
(36) 2×4	(37) 0×5	(38) 2×6	(39) 4×6	(40) 5×2

38

Multiply by 4-6

Time: __________

Score: _____/ 40

(1) 4×4	(2) 5×1	(3) 5×6	(4) 0×6	(5) 5×9
(6) 4×5	(7) 8×5	(8) 3×6	(9) 4×7	(10) 3×5
(11) 6×3	(12) 6×4	(13) 4×6	(14) 0×4	(15) 1×5
(16) 4×9	(17) 6×5	(18) 8×4	(19) 4×2	(20) 2×6
(21) 5×5	(22) 4×1	(23) 6×6	(24) 9×5	(25) 6×8
(26) 3×4	(27) 9×6	(28) 5×4	(29) 2×4	(30) 5×2
(31) 5×3	(32) 4×5	(33) 5×7	(34) 6×7	(35) 5×8
(36) 5×0	(37) 7×4	(38) 5×6	(39) 6×0	(40) 4×4

39

Multiply by 4-6

Time: __________

Score: _____/ 40

(1) 7×4	(2) 5×5	(3) 6×1	(4) 5×4	(5) 3×5
(6) 6×4	(7) 4×5	(8) 1×5	(9) 7×5	(10) 5×6
(11) 9×5	(12) 6×2	(13) 4×9	(14) 5×2	(15) 3×6
(16) 6×7	(17) 6×6	(18) 8×6	(19) 6×9	(20) 4×3
(21) 4×8	(22) 4×6	(23) 8×5	(24) 4×0	(25) 0×5
(26) 6×5	(27) 1×4	(28) 1×6	(29) 6×0	(30) 4×4
(31) 5×9	(32) 5×4	(33) 5×3	(34) 5×0	(35) 4×5
(36) 9×4	(37) 9×6	(38) 6×8	(39) 2×5	(40) 0×4

40

Multiply by 4-6

Time: ________

Score: _____/ 40

(1) 8×4	(2) 8×5	(3) 7×6	(4) 5×5	(5) 6×6
(6) 6×4	(7) 5×6	(8) 1×5	(9) 4×3	(10) 2×6
(11) 7×5	(12) 2×4	(13) 4×2	(14) 6×5	(15) 4×1
(16) 3×6	(17) 4×4	(18) 7×4	(19) 6×0	(20) 4×6
(21) 4×8	(22) 9×4	(23) 4×5	(24) 6×3	(25) 5×7
(26) 4×0	(27) 3×4	(28) 5×6	(29) 5×8	(30) 6×4
(31) 0×6	(32) 6×6	(33) 5×5	(34) 4×7	(35) 2×4
(36) 5×2	(37) 6×7	(38) 6×2	(39) 4×6	(40) 5×0

41

Multiply by 7

Time: ____________

Score: _____/ 40

(1) 3×7	(2) 7×7	(3) 7×1	(4) 7×4	(5) 8×7
(6) 5×7	(7) 7×9	(8) 7×6	(9) 7×2	(10) 0×7
(11) 1×7	(12) 7×7	(13) 9×7	(14) 7×0	(15) 2×7
(16) 7×8	(17) 7×5	(18) 4×7	(19) 7×3	(20) 6×7
(21) 7×4	(22) 7×2	(23) 3×7	(24) 5×7	(25) 7×1
(26) 7×7	(27) 8×7	(28) 0×7	(29) 7×6	(30) 7×9
(31) 7×0	(32) 2×7	(33) 4×7	(34) 7×5	(35) 7×7
(36) 9×7	(37) 1×7	(38) 6×7	(39) 7×8	(40) 7×3

42

Multiply by 7

Time: __________

Score: _____/ 40

① $\begin{array}{r} 0 \\ \times\ 7 \\ \hline \end{array}$	② $\begin{array}{r} 7 \\ \times\ 5 \\ \hline \end{array}$	③ $\begin{array}{r} 6 \\ \times\ 7 \\ \hline \end{array}$	④ $\begin{array}{r} 4 \\ \times\ 7 \\ \hline \end{array}$	⑤ $\begin{array}{r} 7 \\ \times\ 7 \\ \hline \end{array}$
⑥ $\begin{array}{r} 7 \\ \times\ 3 \\ \hline \end{array}$	⑦ $\begin{array}{r} 8 \\ \times\ 7 \\ \hline \end{array}$	⑧ $\begin{array}{r} 7 \\ \times\ 2 \\ \hline \end{array}$	⑨ $\begin{array}{r} 7 \\ \times\ 9 \\ \hline \end{array}$	⑩ $\begin{array}{r} 7 \\ \times\ 1 \\ \hline \end{array}$
⑪ $\begin{array}{r} 7 \\ \times\ 0 \\ \hline \end{array}$	⑫ $\begin{array}{r} 7 \\ \times\ 6 \\ \hline \end{array}$	⑬ $\begin{array}{r} 5 \\ \times\ 7 \\ \hline \end{array}$	⑭ $\begin{array}{r} 1 \\ \times\ 7 \\ \hline \end{array}$	⑮ $\begin{array}{r} 3 \\ \times\ 7 \\ \hline \end{array}$
⑯ $\begin{array}{r} 9 \\ \times\ 7 \\ \hline \end{array}$	⑰ $\begin{array}{r} 7 \\ \times\ 8 \\ \hline \end{array}$	⑱ $\begin{array}{r} 2 \\ \times\ 7 \\ \hline \end{array}$	⑲ $\begin{array}{r} 7 \\ \times\ 4 \\ \hline \end{array}$	⑳ $\begin{array}{r} 7 \\ \times\ 7 \\ \hline \end{array}$
(21) $\begin{array}{r} 7 \\ \times\ 9 \\ \hline \end{array}$	(22) $\begin{array}{r} 4 \\ \times\ 7 \\ \hline \end{array}$	(23) $\begin{array}{r} 8 \\ \times\ 7 \\ \hline \end{array}$	(24) $\begin{array}{r} 7 \\ \times\ 7 \\ \hline \end{array}$	(25) $\begin{array}{r} 7 \\ \times\ 1 \\ \hline \end{array}$
(26) $\begin{array}{r} 7 \\ \times\ 2 \\ \hline \end{array}$	(27) $\begin{array}{r} 0 \\ \times\ 7 \\ \hline \end{array}$	(28) $\begin{array}{r} 7 \\ \times\ 5 \\ \hline \end{array}$	(29) $\begin{array}{r} 6 \\ \times\ 7 \\ \hline \end{array}$	(30) $\begin{array}{r} 7 \\ \times\ 3 \\ \hline \end{array}$
(31) $\begin{array}{r} 7 \\ \times\ 6 \\ \hline \end{array}$	(32) $\begin{array}{r} 3 \\ \times\ 7 \\ \hline \end{array}$	(33) $\begin{array}{r} 5 \\ \times\ 7 \\ \hline \end{array}$	(34) $\begin{array}{r} 7 \\ \times\ 8 \\ \hline \end{array}$	(35) $\begin{array}{r} 7 \\ \times\ 7 \\ \hline \end{array}$
(36) $\begin{array}{r} 9 \\ \times\ 7 \\ \hline \end{array}$	(37) $\begin{array}{r} 7 \\ \times\ 0 \\ \hline \end{array}$	(38) $\begin{array}{r} 7 \\ \times\ 4 \\ \hline \end{array}$	(39) $\begin{array}{r} 2 \\ \times\ 7 \\ \hline \end{array}$	(40) $\begin{array}{r} 1 \\ \times\ 7 \\ \hline \end{array}$

43

Multiply by 7

Time: ________

Score: _____/ 40

(1) 7 × 4	(2) 7 × 5	(3) 7 × 1	(4) 7 × 6	(5) 0 × 7
(6) 7 × 2	(7) 3 × 7	(8) 7 × 7	(9) 8 × 7	(10) 9 × 7
(11) 6 × 7	(12) 7 × 7	(13) 1 × 7	(14) 7 × 0	(15) 2 × 7
(16) 4 × 7	(17) 7 × 8	(18) 5 × 7	(19) 7 × 3	(20) 7 × 9
(21) 7 × 4	(22) 9 × 7	(23) 8 × 7	(24) 0 × 7	(25) 7 × 6
(26) 7 × 1	(27) 7 × 5	(28) 7 × 7	(29) 3 × 7	(30) 7 × 2
(31) 7 × 3	(32) 7 × 0	(33) 7 × 8	(34) 7 × 7	(35) 7 × 9
(36) 5 × 7	(37) 1 × 7	(38) 4 × 7	(39) 6 × 7	(40) 2 × 7

44

Multiply by 7

Time: __________

Score: _____/ 40

(1) 7×1	(2) 7×7	(3) 7×3	(4) 6×7	(5) 7×4
(6) 7×0	(7) 9×7	(8) 7×5	(9) 8×7	(10) 7×2
(11) 0×7	(12) 7×7	(13) 5×7	(14) 2×7	(15) 3×7
(16) 4×7	(17) 7×8	(18) 7×6	(19) 1×7	(20) 7×9
(21) 7×5	(22) 7×1	(23) 7×3	(24) 7×2	(25) 7×0
(26) 7×7	(27) 7×4	(28) 9×7	(29) 8×7	(30) 6×7
(31) 7×9	(32) 5×7	(33) 7×8	(34) 7×7	(35) 3×7
(36) 2×7	(37) 1×7	(38) 4×7	(39) 0×7	(40) 7×6

45

Multiply by 8

Time: ________

Score: _____/ 40

(1) 8×5	(2) 4×8	(3) 8×3	(4) 8×6	(5) 1×8
(6) 8×9	(7) 8×2	(8) 8×8	(9) 8×0	(10) 7×8
(11) 3×8	(12) 8×7	(13) 8×8	(14) 5×8	(15) 8×4
(16) 2×8	(17) 9×8	(18) 0×8	(19) 8×1	(20) 6×8
(21) 8×8	(22) 8×6	(23) 4×8	(24) 8×5	(25) 7×8
(26) 8×9	(27) 8×0	(28) 1×8	(29) 8×2	(30) 8×3
(31) 5×8	(32) 9×8	(33) 8×4	(34) 8×8	(35) 2×8
(36) 6×8	(37) 3×8	(38) 0×8	(39) 8×1	(40) 8×7

46

Multiply by 8

Time: ________

Score: _____/ 40

(1) 9×8	(2) 8×2	(3) 8×1	(4) 8×4	(5) 8×8
(6) 8×0	(7) 5×8	(8) 8×3	(9) 7×8	(10) 8×6
(11) 8×9	(12) 3×8	(13) 1×8	(14) 8×7	(15) 4×8
(16) 6×8	(17) 2×8	(18) 0×8	(19) 8×8	(20) 8×5
(21) 8×8	(22) 8×6	(23) 8×3	(24) 5×8	(25) 8×1
(26) 7×8	(27) 9×8	(28) 8×2	(29) 8×4	(30) 8×0
(31) 2×8	(32) 8×7	(33) 0×8	(34) 1×8	(35) 8×5
(36) 6×8	(37) 8×9	(38) 8×8	(39) 4×8	(40) 3×8

47

Multiply by 8

Time: ________

Score: _____/ 40

(1) 9 × 8	(2) 3 × 8	(3) 8 × 1	(4) 4 × 8	(5) 5 × 8
(6) 8 × 2	(7) 6 × 8	(8) 8 × 8	(9) 8 × 0	(10) 8 × 7
(11) 8 × 5	(12) 0 × 8	(13) 2 × 8	(14) 8 × 8	(15) 8 × 3
(16) 1 × 8	(17) 8 × 4	(18) 7 × 8	(19) 8 × 6	(20) 8 × 9
(21) 5 × 8	(22) 9 × 8	(23) 6 × 8	(24) 8 × 8	(25) 4 × 8
(26) 8 × 2	(27) 8 × 0	(28) 8 × 1	(29) 3 × 8	(30) 8 × 7
(31) 8 × 4	(32) 7 × 8	(33) 8 × 8	(34) 8 × 3	(35) 8 × 6
(36) 2 × 8	(37) 0 × 8	(38) 1 × 8	(39) 8 × 5	(40) 8 × 9

48

Multiply by 8

Time: ________

Score: _____/ 40

(1) 8×8	(2) 1×8	(3) 2×8	(4) 8×0	(5) 6×8
(6) 4×8	(7) 8×5	(8) 8×3	(9) 8×9	(10) 8×7
(11) 0×8	(12) 8×1	(13) 9×8	(14) 8×2	(15) 7×8
(16) 8×6	(17) 5×8	(18) 8×8	(19) 3×8	(20) 8×4
(21) 8×5	(22) 4×8	(23) 8×7	(24) 8×3	(25) 8×0
(26) 8×8	(27) 6×8	(28) 1×8	(29) 8×9	(30) 2×8
(31) 0×8	(32) 8×4	(33) 7×8	(34) 8×2	(35) 8×6
(36) 5×8	(37) 8×1	(38) 8×8	(39) 3×8	(40) 9×8

49

Multiply by 9

Time: ___________

Score: _____/ 40

(1) 9 × 9	(2) 9 × 4	(3) 2 × 9	(4) 9 × 6	(5) 9 × 0
(6) 9 × 8	(7) 9 × 3	(8) 9 × 5	(9) 9 × 1	(10) 9 × 7
(11) 9 × 2	(12) 6 × 9	(13) 3 × 9	(14) 9 × 9	(15) 7 × 9
(16) 0 × 9	(17) 4 × 9	(18) 1 × 9	(19) 8 × 9	(20) 5 × 9
(21) 9 × 4	(22) 9 × 5	(23) 9 × 1	(24) 9 × 8	(25) 9 × 3
(26) 9 × 7	(27) 9 × 0	(28) 9 × 9	(29) 9 × 6	(30) 2 × 9
(31) 9 × 9	(32) 9 × 2	(33) 8 × 9	(34) 7 × 9	(35) 1 × 9
(36) 0 × 9	(37) 6 × 9	(38) 5 × 9	(39) 3 × 9	(40) 4 × 9

50

Multiply by 9

Time: ________

Score: _____/ 40

(1) 9 × 3	(2) 1 × 9	(3) 0 × 9	(4) 7 × 9	(5) 9 × 6
(6) 9 × 9	(7) 5 × 9	(8) 8 × 9	(9) 9 × 2	(10) 9 × 4
(11) 9 × 7	(12) 4 × 9	(13) 9 × 0	(14) 9 × 5	(15) 9 × 9
(16) 2 × 9	(17) 9 × 1	(18) 9 × 8	(19) 3 × 9	(20) 6 × 9
(21) 1 × 9	(22) 9 × 9	(23) 5 × 9	(24) 7 × 9	(25) 9 × 6
(26) 9 × 3	(27) 8 × 9	(28) 0 × 9	(29) 9 × 4	(30) 9 × 2
(31) 9 × 1	(32) 9 × 5	(33) 4 × 9	(34) 9 × 7	(35) 2 × 9
(36) 3 × 9	(37) 9 × 8	(38) 9 × 9	(39) 9 × 0	(40) 6 × 9

51

Multiply by 9

Time: ________

Score: _____/ 40

(1) 4 × 9	(2) 8 × 9	(3) 0 × 9	(4) 9 × 9	(5) 9 × 7
(6) 2 × 9	(7) 9 × 1	(8) 9 × 6	(9) 5 × 9	(10) 3 × 9
(11) 9 × 8	(12) 9 × 2	(13) 9 × 9	(14) 9 × 0	(15) 9 × 5
(16) 7 × 9	(17) 1 × 9	(18) 6 × 9	(19) 9 × 3	(20) 9 × 4
(21) 2 × 9	(22) 5 × 9	(23) 9 × 1	(24) 9 × 7	(25) 9 × 9
(26) 9 × 6	(27) 4 × 9	(28) 8 × 9	(29) 0 × 9	(30) 3 × 9
(31) 6 × 9	(32) 9 × 8	(33) 9 × 2	(34) 9 × 4	(35) 9 × 3
(36) 9 × 9	(37) 7 × 9	(38) 9 × 5	(39) 1 × 9	(40) 9 × 0

52

Multiply by 9

Time: ________

Score: _____/ 40

① 7×9	② 4×9	③ 9×1	④ 9×3	⑤ 9×6
⑥ 9×5	⑦ 2×9	⑧ 9×0	⑨ 9×9	⑩ 9×8
⑪ 1×9	⑫ 8×9	⑬ 9×7	⑭ 0×9	⑮ 9×4
⑯ 9×9	⑰ 3×9	⑱ 5×9	⑲ 9×2	⑳ 6×9
(21) 7×9	(22) 9×6	(23) 9×1	(24) 9×3	(25) 9×5
(26) 2×9	(27) 9×0	(28) 9×8	(29) 9×9	(30) 4×9
(31) 1×9	(32) 6×9	(33) 8×9	(34) 3×9	(35) 9×7
(36) 5×9	(37) 9×4	(38) 0×9	(39) 9×2	(40) 9×9

53

Multiply by 7-9

Time: ___________

Score: _____/ 40

(1) 9×7	(2) 4×9	(3) 10×9	(4) 9×9	(5) 6×9
(6) 7×3	(7) 7×9	(8) 1×9	(9) 7×5	(10) 7×8
(11) 7×4	(12) 2×8	(13) 7×0	(14) 9×8	(15) 9×0
(16) 8×1	(17) 8×6	(18) 7×10	(19) 10×8	(20) 9×2
(21) 7×7	(22) 8×7	(23) 0×8	(24) 4×8	(25) 3×9
(26) 2×7	(27) 8×3	(28) 5×9	(29) 6×7	(30) 5×8
(31) 8×9	(32) 8×8	(33) 1×7	(34) 1×8	(35) 8×0
(36) 3×7	(37) 8×8	(38) 9×3	(39) 8×10	(40) 7×1

54

Multiply by 7-9

Time: ________

Score: _____/ 40

① 4 × 9	② 5 × 8	③ 6 × 7	④ 7 × 8	⑤ 4 × 8
⑥ 9 × 10	⑦ 8 × 6	⑧ 9 × 1	⑨ 5 × 9	⑩ 7 × 0
⑪ 5 × 7	⑫ 3 × 8	⑬ 8 × 9	⑭ 2 × 7	⑮ 9 × 8
⑯ 9 × 2	⑰ 7 × 4	⑱ 9 × 7	⑲ 7 × 9	⑳ 0 × 9
(21) 2 × 8	(22) 6 × 9	(23) 7 × 7	(24) 9 × 9	(25) 8 × 7
(26) 10 × 7	(27) 0 × 7	(28) 7 × 7	(29) 8 × 1	(30) 8 × 0
(31) 9 × 7	(32) 8 × 5	(33) 9 × 6	(34) 7 × 1	(35) 7 × 8
(36) 8 × 3	(37) 1 × 9	(38) 6 × 8	(39) 10 × 8	(40) 8 × 2

55

Multiply by 7-9

Time: ________

Score: _____/ 40

(1) 9×4	(2) 5×7	(3) 5×9	(4) 8×4	(5) 8×9
(6) 9×9	(7) 4×7	(8) 7×6	(9) 7×2	(10) 9×10
(11) 8×8	(12) 7×8	(13) 7×9	(14) 9×0	(15) 9×8
(16) 9×3	(17) 2×9	(18) 3×7	(19) 10×7	(20) 9×9
(21) 9×7	(22) 7×9	(23) 5×8	(24) 8×7	(25) 9×2
(26) 8×9	(27) 3×9	(28) 9×5	(29) 1×9	(30) 8×1
(31) 8×8	(32) 0×9	(33) 7×10	(34) 7×3	(35) 9×8
(36) 2×8	(37) 7×8	(38) 8×10	(39) 7×0	(40) 6×9

56

Multiply by 7-9

Time: ________

Score: _____/ 40

(1) 7 × 5	(2) 10 × 9	(3) 6 × 7	(4) 8 × 6	(5) 3 × 8
(6) 4 × 8	(7) 0 × 8	(8) 7 × 7	(9) 2 × 7	(10) 1 × 7
(11) 9 × 4	(12) 4 × 7	(13) 2 × 8	(14) 1 × 8	(15) 7 × 6
(16) 8 × 4	(17) 4 × 9	(18) 7 × 9	(19) 7 × 1	(20) 9 × 1
(21) 9 × 10	(22) 9 × 2	(23) 7 × 3	(24) 9 × 9	(25) 5 × 7
(26) 6 × 9	(27) 0 × 7	(28) 7 × 7	(29) 7 × 8	(30) 8 × 9
(31) 6 × 8	(32) 8 × 3	(33) 9 × 0	(34) 8 × 0	(35) 3 × 9
(36) 10 × 7	(37) 8 × 8	(38) 9 × 8	(39) 7 × 2	(40) 9 × 7

57

Multiply by 7-9

Time: ________

Score: _____/ 40

① 7×8	② 5×8	③ 8×10	④ 5×9	⑤ 4×7
⑥ 2×9	⑦ 9×3	⑧ 8×5	⑨ 0×8	⑩ 5×7
⑪ 9×4	⑫ 8×8	⑬ 9×5	⑭ 6×7	⑮ 7×9
⑯ 1×8	⑰ 10×8	⑱ 8×4	⑲ 10×9	⑳ 8×7
(21) 7×1	(22) 3×8	(23) 9×9	(24) 9×1	(25) 10×7
(26) 7×8	(27) 9×8	(28) 8×9	(29) 3×7	(30) 6×8
(31) 8×2	(32) 9×7	(33) 7×7	(34) 7×0	(35) 7×4
(36) 9×0	(37) 9×6	(38) 2×7	(39) 8×10	(40) 5×9

58

Multiply by 7-9

Time: __________

Score: _____/ 40

① 8×3	② 9×2	③ 7×8	④ 7×7	⑤ 2×8
⑥ 8×7	⑦ 7×6	⑧ 8×5	⑨ 9×6	⑩ 7×3
⑪ 7×9	⑫ 9×9	⑬ 10×7	⑭ 5×7	⑮ 9×0
⑯ 0×8	⑰ 8×9	⑱ 7×4	⑲ 7×1	⑳ 1×9
(21) 8×8	(22) 9×3	(23) 4×9	(24) 9×8	(25) 6×8
(26) 0×7	(27) 10×9	(28) 9×7	(29) 7×2	(30) 1×8
(31) 8×4	(32) 8×9	(33) 10×8	(34) 0×9	(35) 1×7
(36) 9×10	(37) 3×9	(38) 8×1	(39) 9×4	(40) 7×9

59

Multiply by 7-9

Time: ____________

Score: ______/ 40

(1) 3×7	(2) 7×8	(3) 2×9	(4) 6×9	(5) 6×7
(6) 8×6	(7) 1×9	(8) 8×7	(9) 2×8	(10) 8×8
(11) 0×7	(12) 8×0	(13) 7×7	(14) 7×4	(15) 9×9
(16) 7×10	(17) 3×8	(18) 7×2	(19) 9×5	(20) 7×5
(21) 8×4	(22) 5×8	(23) 9×8	(24) 9×7	(25) 8×5
(26) 3×9	(27) 9×9	(28) 9×1	(29) 9×6	(30) 7×9
(31) 0×8	(32) 8×9	(33) 7×6	(34) 8×10	(35) 4×7
(36) 5×9	(37) 7×7	(38) 4×8	(39) 7×8	(40) 9×8

60

Multiply by 7-9

Time: __________

Score: _____/ 40

① 8 × 7	② 5 × 7	③ 9 × 7	④ 7 × 10	⑤ 9 × 2
⑥ 2 × 8	⑦ 8 × 8	⑧ 9 × 4	⑨ 9 × 10	⑩ 9 × 0
⑪ 1 × 7	⑫ 8 × 1	⑬ 3 × 7	⑭ 2 × 7	⑮ 0 × 7
⑯ 8 × 6	⑰ 3 × 8	⑱ 8 × 8	⑲ 7 × 9	⑳ 9 × 9
21 2 × 9	22 8 × 5	23 7 × 5	24 0 × 9	25 7 × 1
26 8 × 3	27 10 × 9	28 9 × 7	29 7 × 2	30 5 × 9
31 4 × 8	32 4 × 7	33 6 × 8	34 1 × 9	35 8 × 2
36 10 × 7	37 7 × 8	38 8 × 7	39 9 × 6	40 3 × 9

61

Multiply by 10

Time: ________

Score: _____/ 40

(1) 10×4	(2) 10×8	(3) 9×10	(4) 10×0	(5) 3×10
(6) 1×10	(7) 10×10	(8) 6×10	(9) 10×5	(10) 10×2
(11) 10×7	(12) 5×10	(13) 2×10	(14) 10×9	(15) 4×10
(16) 0×10	(17) 10×3	(18) 8×10	(19) 7×10	(20) 10×6
(21) 10×10	(22) 10×1	(23) 10×4	(24) 10×7	(25) 6×10
(26) 9×10	(27) 10×2	(28) 10×5	(29) 1×10	(30) 10×10
(31) 3×10	(32) 10×8	(33) 10×0	(34) 10×9	(35) 0×10
(36) 4×10	(37) 8×10	(38) 10×1	(39) 5×10	(40) 2×10

62

Multiply by 10

Time: ________

Score: _____/ 40

1. 3×10
2. 6×10
3. 10×10
4. 7×10
5. 0×10
6. 10×6
7. 10×9
8. 10×4
9. 10×3
10. 5×10
11. 10×7
12. 10×8
13. 10×10
14. 1×10
15. 2×10
16. 10×10
17. 10×2
18. 9×10
19. 7×10
20. 6×10
21. 8×10
22. 10×5
23. 10×0
24. 4×10
25. 10×1
26. 3×10
27. 10×7
28. 10×10
29. 10×3
30. 1×10
31. 10×8
32. 10×4
33. 5×10
34. 10×9
35. 10×6
36. 0×10
37. 2×10
38. 10×0
39. 10×5
40. 10×10

63

Multiply by 10

Time: ________

Score: _____/ 40

(1) 10 × 8	(2) 10 × 3	(3) 10 × 7	(4) 10 × 6	(5) 9 × 10
(6) 4 × 10	(7) 2 × 10	(8) 10 × 1	(9) 10 × 2	(10) 10 × 4
(11) 3 × 10	(12) 5 × 10	(13) 10 × 10	(14) 6 × 10	(15) 7 × 10
(16) 8 × 10	(17) 0 × 10	(18) 10 × 9	(19) 1 × 10	(20) 10 × 6
(21) 4 × 10	(22) 10 × 8	(23) 10 × 5	(24) 2 × 10	(25) 10 × 0
(26) 10 × 7	(27) 9 × 10	(28) 10 × 10	(29) 10 × 1	(30) 10 × 3
(31) 1 × 10	(32) 10 × 9	(33) 8 × 10	(34) 7 × 10	(35) 10 × 10
(36) 10 × 4	(37) 10 × 2	(38) 0 × 10	(39) 5 × 10	(40) 3 × 10

64

Multiply by 10

Time: ____________

Score: _____/ 40

(1) 6×10	(2) 10×9	(3) 10×6	(4) 3×10	(5) 10×4
(6) 10×8	(7) 10×5	(8) 10×7	(9) 10×10	(10) 10×2
(11) 0×10	(12) 1×10	(13) 10×0	(14) 8×10	(15) 10×1
(16) 10×10	(17) 5×10	(18) 10×3	(19) 6×10	(20) 7×10
(21) 4×10	(22) 2×10	(23) 9×10	(24) 10×7	(25) 10×5
(26) 10×4	(27) 10×10	(28) 10×6	(29) 1×10	(30) 10×2
(31) 0×10	(32) 3×10	(33) 10×9	(34) 10×8	(35) 5×10
(36) 2×10	(37) 6×10	(38) 10×10	(39) 8×10	(40) 10×0

65

Multiply by 11

Time: ________

Score: _____/ 40

(1) 11×5	(2) 1×11	(3) 11×7	(4) 11×8	(5) 11×6
(6) 11×2	(7) 0×11	(8) 11×4	(9) 11×9	(10) 3×11
(11) 11×10	(12) 11×11	(13) 11×3	(14) 5×11	(15) 11×0
(16) 4×11	(17) 10×11	(18) 8×11	(19) 2×11	(20) 6×11
(21) 9×11	(22) 11×1	(23) 11×11	(24) 7×11	(25) 11×4
(26) 11×2	(27) 11×10	(28) 11×7	(29) 11×8	(30) 3×11
(31) 1×11	(32) 11×9	(33) 0×11	(34) 11×11	(35) 11×6
(36) 11×5	(37) 2×11	(38) 5×11	(39) 4×11	(40) 11×3

66

Multiply by 11

Time: __________

Score: _____/ 40

① 11×9	② 6×11	③ 11×7	④ 11×8	⑤ 10×11
⑥ 11×11	⑦ 11×1	⑧ 11×0	⑨ 5×11	⑩ 11×11
⑪ 11×3	⑫ 2×11	⑬ 9×11	⑭ 7×11	⑮ 4×11
⑯ 0×11	⑰ 11×10	⑱ 11×6	⑲ 1×11	⑳ 8×11
(21) 11×4	(22) 11×5	(23) 3×11	(24) 6×11	(25) 10×11
(26) 11×9	(27) 11×7	(28) 11×1	(29) 11×8	(30) 11×0
(31) 11×11	(32) 11×2	(33) 1×11	(34) 0×11	(35) 2×11
(36) 4×11	(37) 5×11	(38) 11×3	(39) 11×11	(40) 9×11

67

Multiply by 11

Time: ____________

Score: ______/ 40

(1) 11×8	(2) 11×6	(3) 11×7	(4) 11×10	(5) 6×11
(6) 11×2	(7) 11×9	(8) 10×11	(9) 11×0	(10) 5×11
(11) 8×11	(12) 4×11	(13) 11×3	(14) 11×1	(15) 11×11
(16) 7×11	(17) 3×11	(18) 11×6	(19) 1×11	(20) 11×10
(21) 11×11	(22) 9×11	(23) 11×7	(24) 2×11	(25) 11×8
(26) 11×5	(27) 11×4	(28) 0×11	(29) 11×1	(30) 11×0
(31) 6×11	(32) 11×11	(33) 4×11	(34) 5×11	(35) 11×2
(36) 7×11	(37) 11×9	(38) 11×3	(39) 10×11	(40) 8×11

68

Multiply by 11

Time: ________

Score: _____/ 40

(1) 11×11	(2) 9×11	(3) 11×2	(4) 11×6	(5) 8×11
(6) 5×11	(7) 11×7	(8) 11×1	(9) 10×11	(10) 11×0
(11) 4×11	(12) 3×11	(13) 2×11	(14) 6×11	(15) 11×10
(16) 11×4	(17) 0×11	(18) 7×11	(19) 11×11	(20) 11×9
(21) 11×3	(22) 1×11	(23) 11×8	(24) 11×5	(25) 11×2
(26) 10×11	(27) 5×11	(28) 4×11	(29) 11×0	(30) 11×1
(31) 11×11	(32) 9×11	(33) 11×6	(34) 8×11	(35) 11×7
(36) 3×11	(37) 11×10	(38) 1×11	(39) 11×5	(40) 11×8

69

Multiply by 12

Time: ________

Score: _____/ 40

① 12 × 6	② 12 × 9	③ 4 × 12	④ 12 × 10	⑤ 11 × 12
⑥ 12 × 1	⑦ 12 × 12	⑧ 8 × 12	⑨ 12 × 5	⑩ 12 × 0
⑪ 7 × 12	⑫ 2 × 12	⑬ 3 × 12	⑭ 12 × 2	⑮ 9 × 12
⑯ 10 × 12	⑰ 12 × 3	⑱ 12 × 8	⑲ 12 × 12	⑳ 12 × 11
(21) 5 × 12	(22) 12 × 7	(23) 6 × 12	(24) 0 × 12	(25) 1 × 12
(26) 12 × 4	(27) 8 × 12	(28) 4 × 12	(29) 12 × 6	(30) 12 × 9
(31) 11 × 12	(32) 12 × 0	(33) 2 × 12	(34) 12 × 5	(35) 3 × 12
(36) 12 × 1	(37) 12 × 12	(38) 12 × 10	(39) 7 × 12	(40) 1 × 12

70

Multiply by 12

Time: __________

Score: _____/ 40

1. 12×7	2. 12×2	3. 12×0	4. 12×10	5. 12×11
6. 12×6	7. 12×12	8. 4×12	9. 12×9	10. 12×3
11. 5×12	12. 8×12	13. 3×12	14. 2×12	15. 12×1
16. 6×12	17. 10×12	18. 7×12	19. 12×5	20. 0×12
21. 11×12	22. 12×8	23. 9×12	24. 12×4	25. 12×12
26. 4×12	27. 12×6	28. 12×7	29. 12×3	30. 12×11
31. 1×12	32. 12×10	33. 12×12	34. 12×2	35. 5×12
36. 12×0	37. 8×12	38. 12×9	39. 2×12	40. 12×1

71

Multiply by 12

Time: ________

Score: ______/ 40

(1) 12×10	(2) 0×12	(3) 6×12	(4) 7×12	(5) 8×12
(6) 12×12	(7) 12×11	(8) 4×12	(9) 12×5	(10) 3×12
(11) 12×9	(12) 11×12	(13) 9×12	(14) 12×4	(15) 12×8
(16) 10×12	(17) 2×12	(18) 12×12	(19) 1×12	(20) 12×7
(21) 12×0	(22) 12×6	(23) 5×12	(24) 12×3	(25) 12×5
(26) 12×12	(27) 0×12	(28) 12×10	(29) 3×12	(30) 8×12
(31) 4×12	(32) 12×9	(33) 12×11	(34) 6×12	(35) 7×12
(36) 12×1	(37) 12×2	(38) 11×12	(39) 9×12	(40) 10×12

72

Multiply by 12

Time: ________

Score: _____/ 40

① 8 × 12	② 12 × 4	③ 1 × 12	④ 12 × 12	⑤ 2 × 12
⑥ 12 × 6	⑦ 7 × 12	⑧ 12 × 3	⑨ 12 × 0	⑩ 5 × 12
⑪ 12 × 11	⑫ 12 × 2	⑬ 6 × 12	⑭ 12 × 9	⑮ 12 × 12
⑯ 12 × 7	⑰ 12 × 8	⑱ 4 × 12	⑲ 3 × 12	⑳ 12 × 1
(21) 12 × 10	(22) 12 × 5	(23) 0 × 12	(24) 12 × 3	(25) 1 × 12
(26) 12 × 12	(27) 7 × 12	(28) 11 × 12	(29) 12 × 0	(30) 5 × 12
(31) 12 × 4	(32) 2 × 12	(33) 9 × 12	(34) 8 × 12	(35) 12 × 6
(36) 10 × 12	(37) 3 × 12	(38) 12 × 10	(39) 12 × 2	(40) 0 × 12

73

Multiply by 10-12

Time: ________

Score: _____/ 40

(1) 3×11	(2) 12×11	(3) 12×7	(4) 10×9	(5) 12×10
(6) 10×11	(7) 4×12	(8) 1×12	(9) 10×8	(10) 7×11
(11) 10×10	(12) 6×12	(13) 9×11	(14) 12×9	(15) 0×10
(16) 7×10	(17) 11×11	(18) 2×12	(19) 3×10	(20) 12×5
(21) 12×12	(22) 3×12	(23) 10×2	(24) 12×4	(25) 12×0
(26) 6×10	(27) 11×12	(28) 11×2	(29) 10×12	(30) 1×11
(31) 5×11	(32) 11×0	(33) 11×6	(34) 8×12	(35) 11×10
(36) 11×8	(37) 10×1	(38) 11×4	(39) 10×4	(40) 8×11

74

Multiply by 10-12

Time: ________

Score: _____/ 40

① 4×10	② 1×12	③ 12×8	④ 7×11	⑤ 6×12
⑥ 11×12	⑦ 4×12	⑧ 12×4	⑨ 12×3	⑩ 12×7
⑪ 10×9	⑫ 10×10	⑬ 7×10	⑭ 12×0	⑮ 11×11
⑯ 12×12	⑰ 11×0	⑱ 11×10	⑲ 10×1	⑳ 9×11
(21) 10×12	(22) 5×11	(23) 11×3	(24) 10×3	(25) 12×11
(26) 11×4	(27) 8×10	(28) 10×0	(29) 11×1	(30) 6×10
(31) 2×12	(32) 6×11	(33) 9×12	(34) 10×2	(35) 12×10
(36) 2×11	(37) 10×11	(38) 12×5	(39) 10×6	(40) 10×10

75

Multiply by 10-12

Time: ________

Score: _____/ 40

① 3 × 11	② 11 × 2	③ 11 × 9	④ 12 × 12	⑤ 12 × 4
⑥ 11 × 0	⑦ 12 × 11	⑧ 12 × 3	⑨ 12 × 10	⑩ 11 × 5
⑪ 1 × 11	⑫ 4 × 11	⑬ 10 × 4	⑭ 11 × 12	⑮ 8 × 10
⑯ 7 × 12	⑰ 8 × 12	⑱ 11 × 6	⑲ 4 × 12	⑳ 12 × 6
(21) 11 × 8	(22) 0 × 10	(23) 3 × 10	(24) 12 × 9	(25) 10 × 12
(26) 10 × 9	(27) 1 × 12	(28) 5 × 12	(29) 11 × 11	(30) 10 × 1
(31) 10 × 11	(32) 12 × 0	(33) 2 × 12	(34) 7 × 10	(35) 10 × 2
(36) 11 × 10	(37) 11 × 7	(38) 10 × 7	(39) 1 × 10	(40) 4 × 10

76

Multiply by 10-12

Time: ___________

Score: _____/ 40

(1) 9×10	(2) 7×11	(3) 11×0	(4) 5×12	(5) 2×11
(6) 10×11	(7) 11×3	(8) 8×10	(9) 10×2	(10) 12×0
(11) 4×12	(12) 11×11	(13) 11×9	(14) 10×3	(15) 3×12
(16) 10×6	(17) 10×10	(18) 12×4	(19) 11×10	(20) 12×2
(21) 12×11	(22) 8×12	(23) 5×11	(24) 0×10	(25) 6×12
(26) 9×12	(27) 12×7	(28) 12×12	(29) 12×10	(30) 6×11
(31) 11×12	(32) 4×11	(33) 12×1	(34) 1×11	(35) 10×12
(36) 8×11	(37) 11×1	(38) 10×9	(39) 10×11	(40) 1×12

77

Multiply by 10-12

Time: ________

Score: _____/ 40

(1) 11 × 5	(2) 12 × 11	(3) 2 × 11	(4) 0 × 10	(5) 11 × 6
(6) 8 × 11	(7) 12 × 4	(8) 4 × 10	(9) 12 × 8	(10) 10 × 1
(11) 10 × 2	(12) 4 × 12	(13) 11 × 0	(14) 7 × 11	(15) 3 × 12
(16) 7 × 10	(17) 12 × 7	(18) 3 × 10	(19) 11 × 11	(20) 9 × 11
(21) 6 × 12	(22) 12 × 12	(23) 6 × 10	(24) 0 × 12	(25) 12 × 10
(26) 11 × 12	(27) 3 × 11	(28) 2 × 12	(29) 10 × 10	(30) 4 × 11
(31) 10 × 8	(32) 10 × 12	(33) 5 × 12	(34) 12 × 9	(35) 10 × 11
(36) 12 × 3	(37) 11 × 10	(38) 1 × 11	(39) 12 × 5	(40) 12 × 11

78

Multiply by 10-12

Time: ________

Score: _____/ 40

(1) 10 × 1	(2) 10 × 6	(3) 9 × 10	(4) 11 × 11	(5) 12 × 0
(6) 2 × 10	(7) 4 × 12	(8) 2 × 12	(9) 12 × 4	(10) 11 × 10
(11) 11 × 8	(12) 10 × 10	(13) 11 × 3	(14) 11 × 4	(15) 12 × 10
(16) 12 × 1	(17) 11 × 2	(18) 12 × 12	(19) 4 × 10	(20) 11 × 6
(21) 11 × 5	(22) 8 × 12	(23) 7 × 10	(24) 3 × 10	(25) 11 × 12
(26) 10 × 12	(27) 11 × 9	(28) 12 × 6	(29) 11 × 7	(30) 12 × 9
(31) 10 × 0	(32) 12 × 7	(33) 10 × 8	(34) 0 × 11	(35) 7 × 11
(36) 0 × 12	(37) 12 × 8	(38) 10 × 2	(39) 6 × 10	(40) 11 × 0

79

Multiply by 10-12

Time: ________

Score: _____/ 40

(1) 9 × 10	(2) 2 × 12	(3) 10 × 11	(4) 1 × 12	(5) 11 × 4
(6) 11 × 12	(7) 11 × 1	(8) 9 × 11	(9) 3 × 11	(10) 2 × 11
(11) 11 × 6	(12) 12 × 7	(13) 10 × 5	(14) 10 × 7	(15) 10 × 0
(16) 12 × 3	(17) 5 × 10	(18) 6 × 12	(19) 11 × 11	(20) 9 × 12
(21) 8 × 11	(22) 12 × 10	(23) 10 × 10	(24) 10 × 12	(25) 3 × 10
(26) 1 × 10	(27) 10 × 8	(28) 11 × 10	(29) 12 × 5	(30) 12 × 11
(31) 4 × 10	(32) 12 × 12	(33) 5 × 11	(34) 5 × 12	(35) 6 × 11
(36) 11 × 5	(37) 6 × 10	(38) 12 × 6	(39) 7 × 12	(40) 10 × 1

80
Multiply by 10-12

Time: ________

Score: _____/ 40

(1) 1 × 11	(2) 10 × 9	(3) 12 × 1	(4) 2 × 12	(5) 8 × 11
(6) 10 × 2	(7) 5 × 10	(8) 9 × 11	(9) 8 × 12	(10) 11 × 7
(11) 11 × 10	(12) 10 × 10	(13) 10 × 4	(14) 12 × 0	(15) 10 × 7
(16) 11 × 3	(17) 11 × 11	(18) 12 × 12	(19) 11 × 4	(20) 11 × 0
(21) 0 × 10	(22) 10 × 11	(23) 3 × 12	(24) 11 × 12	(25) 3 × 10
(26) 10 × 12	(27) 8 × 10	(28) 9 × 12	(29) 12 × 11	(30) 10 × 5
(31) 11 × 2	(32) 12 × 10	(33) 12 × 12	(34) 12 × 8	(35) 10 × 8
(36) 12 × 3	(37) 10 × 6	(38) 5 × 10	(39) 10 × 12	(40) 11 × 12

81

Multiply by 0-12

Time: ________

Score: _____/ 40

(1) 0×11	(2) 3×12	(3) 2×0	(4) 7×12	(5) 10×10
(6) 7×4	(7) 8×11	(8) 2×12	(9) 12×10	(10) 1×9
(11) 10×12	(12) 11×8	(13) 6×12	(14) 12×3	(15) 6×8
(16) 0×0	(17) 11×5	(18) 11×1	(19) 0×4	(20) 3×12
(21) 7×0	(22) 12×5	(23) 2×4	(24) 3×10	(25) 7×2
(26) 9×5	(27) 7×8	(28) 3×2	(29) 1×7	(30) 0×3
(31) 10×3	(32) 12×2	(33) 6×9	(34) 6×11	(35) 7×10
(36) 12×12	(37) 11×10	(38) 0×7	(39) 12×1	(40) 7×5

82

Multiply by 0-12

Time: ________

Score: _____/ 40

(1) 5×8	(2) 9×7	(3) 5×3	(4) 4×4	(5) 5×0
(6) 11×11	(7) 7×0	(8) 9×11	(9) 5×11	(10) 8×2
(11) 2×2	(12) 1×3	(13) 4×3	(14) 9×4	(15) 3×2
(16) 3×11	(17) 9×3	(18) 8×8	(19) 1×2	(20) 3×9
(21) 2×11	(22) 3×7	(23) 10×2	(24) 3×4	(25) 5×7
(26) 8×12	(27) 6×1	(28) 0×2	(29) 4×4	(30) 5×9
(31) 1×12	(32) 3×5	(33) 3×3	(34) 7×4	(35) 6×5
(36) 2×8	(37) 10×7	(38) 1×6	(39) 6×7	(40) 12×11

83
Multiply by 0-12

Time: ________

Score: _____/ 40

1. 8×3
2. 1×1
3. 4×0
4. 2×5
5. 4×12
6. 8×2
7. 1×3
8. 5×1
9. 12×6
10. 10×6
11. 5×4
12. 10×4
13. 2×6
14. 4×10
15. 4×6
16. 4×8
17. 12×4
18. 7×11
19. 0×11
20. 8×12
21. 1×0
22. 10×0
23. 6×4
24. 2×8
25. 0×10
26. 10×11
27. 10×8
28. 7×3
29. 4×11
30. 10×9
31. 8×0
32. 11×7
33. 9×10
34. 1×4
35. 9×8
36. 9×1
37. 4×9
38. 7×9
39. 0×1
40. 6×0

84

Multiply by 0-12

Time: __________

Score: _____/ 40

(1) 5×6 = ___	(2) 1×8 = ___	(3) 12×11 = ___	(4) 3×11 = ___	(5) 12×5 = ___
(6) 2×10 = ___	(7) 6×0 = ___	(8) 12×9 = ___	(9) 5×5 = ___	(10) 0×12 = ___
(11) 8×0 = ___	(12) 4×8 = ___	(13) 7×8 = ___	(14) 6×6 = ___	(15) 7×7 = ___
(16) 7×6 = ___	(17) 9×9 = ___	(18) 7×1 = ___	(19) 10×5 = ___	(20) 8×9 = ___
(21) 5×10 = ___	(22) 2×11 = ___	(23) 12×0 = ___	(24) 9×2 = ___	(25) 6×8 = ___
(26) 5×0 = ___	(27) 8×3 = ___	(28) 0×12 = ___	(29) 10×6 = ___	(30) 9×11 = ___
(31) 8×10 = ___	(32) 1×2 = ___	(33) 5×1 = ___	(34) 11×1 = ___	(35) 9×0 = ___
(36) 8×1 = ___	(37) 9×12 = ___	(38) 10×1 = ___	(39) 12×7 = ___	(40) 0×10 = ___

85

Multiply by 0-12

Time: __________

Score: _____/ 40

(1) 2 × 6	(2) 9 × 0	(3) 11 × 4	(4) 9 × 2	(5) 5 × 4
(6) 0 × 3	(7) 10 × 1	(8) 9 × 6	(9) 6 × 11	(10) 3 × 11
(11) 10 × 3	(12) 10 × 5	(13) 11 × 0	(14) 8 × 6	(15) 12 × 2
(16) 11 × 8	(17) 12 × 0	(18) 0 × 8	(19) 5 × 12	(20) 4 × 5
(21) 11 × 5	(22) 7 × 1	(23) 1 × 8	(24) 5 × 10	(25) 12 × 11
(26) 1 × 1	(27) 8 × 7	(28) 2 × 1	(29) 6 × 6	(30) 4 × 2
(31) 12 × 7	(32) 1 × 2	(33) 8 × 0	(34) 4 × 1	(35) 3 × 6
(36) 6 × 3	(37) 1 × 3	(38) 9 × 3	(39) 1 × 10	(40) 1 × 7

86

Multiply by 0-12

Time: ________

Score: ______/ 40

1. 9×6	2. 6×4	3. 10×8	4. 5×7	5. 4×6
6. 2×5	7. 6×10	8. 1×6	9. 4×11	10. 6×4
11. 9×9	12. 6×5	13. 1×9	14. 6×1	15. 8×12
16. 5×2	17. 2×3	18. 2×2	19. 0×2	20. 2×11
21. 5×0	22. 6×2	23. 9×10	24. 5×9	25. 10×3
26. 11×10	27. 2×5	28. 12×3	29. 12×6	30. 9×8
31. 1×3	32. 6×9	33. 1×0	34. 2×2	35. 1×6
36. 0×7	37. 9×7	38. 9×4	39. 3×10	40. 4×4

87

Multiply by 0-12

Time: ____________

Score: _____/ 40

(1) 8×0	(2) 8×12	(3) 10×6	(4) 11×2	(5) 3×7
(6) 11×10	(7) 3×4	(8) 0×2	(9) 4×8	(10) 9×11
(11) 5×12	(12) 1×12	(13) 5×3	(14) 7×3	(15) 8×5
(16) 5×8	(17) 10×12	(18) 9×4	(19) 5×0	(20) 9×7
(21) 11×11	(22) 10×10	(23) 1×10	(24) 3×8	(25) 12×1
(26) 3×11	(27) 0×0	(28) 10×8	(29) 0×11	(30) 10×0
(31) 5×5	(32) 9×3	(33) 11×9	(34) 12×4	(35) 5×7
(36) 10×2	(37) 10×4	(38) 12×7	(39) 12×10	(40) 3×5

88

Multiply by 0-12

Time: ____________

Score: _____/ 40

(1) 12×12	(2) 12×6	(3) 8×8	(4) 11×6	(5) 8×2
(6) 4×3	(7) 4×1	(8) 2×12	(9) 12×2	(10) 4×10
(11) 1×11	(12) 1×5	(13) 1×2	(14) 8×6	(15) 1×0
(16) 3×3	(17) 5×4	(18) 9×1	(19) 10×9	(20) 3×0
(21) 0×9	(22) 3×8	(23) 7×11	(24) 6×11	(25) 11×5
(26) 2×9	(27) 10×4	(28) 1×4	(29) 9×8	(30) 7×8
(31) 7×7	(32) 0×6	(33) 2×1	(34) 4×10	(35) 7×2
(36) 11×7	(37) 3×4	(38) 9×2	(39) 12×9	(40) 5×8

89

Multiply by 0-12

Time: __________

Score: _____/ 40

(1) 11 × 8	(2) 12 × 9	(3) 7 × 2	(4) 5 × 10	(5) 4 × 11
(6) 5 × 9	(7) 4 × 7	(8) 2 × 6	(9) 1 × 6	(10) 6 × 3
(11) 12 × 11	(12) 1 × 5	(13) 7 × 4	(14) 0 × 9	(15) 10 × 7
(16) 7 × 10	(17) 1 × 11	(18) 3 × 0	(19) 4 × 7	(20) 6 × 11
(21) 4 × 6	(22) 2 × 11	(23) 5 × 5	(24) 8 × 6	(25) 12 × 0
(26) 7 × 9	(27) 6 × 7	(28) 7 × 8	(29) 0 × 3	(30) 2 × 7
(31) 1 × 2	(32) 2 × 9	(33) 11 × 1	(34) 12 × 6	(35) 8 × 7
(36) 8 × 0	(37) 7 × 6	(38) 6 × 10	(39) 0 × 12	(40) 11 × 12

90

Multiply by 0-12

Time: __________

Score: _____/ 40

① 10×10	② 10×11	③ 3×5	④ 5×1	⑤ 0×2
⑥ 4×6	⑦ 2×3	⑧ 5×0	⑨ 7×0	⑩ 6×6
⑪ 2×5	⑫ 6×0	⑬ 3×12	⑭ 8×3	⑮ 0×1
⑯ 10×7	⑰ 12×3	⑱ 9×5	⑲ 4×9	⑳ 3×6
(21) 1×1	(22) 10×0	(23) 10×9	(24) 11×3	(25) 7×2
(26) 5×12	(27) 9×10	(28) 10×6	(29) 3×8	(30) 11×5
(31) 2×7	(32) 12×7	(33) 5×4	(34) 7×1	(35) 12×4
(36) 11×9	(37) 8×1	(38) 6×12	(39) 12×12	(40) 4×8

91

Multiply by 0-12

Time: ____________

Score: ______/ 40

(1) 8×2	(2) 2×4	(3) 4×2	(4) 3×3	(5) 0×5
(6) 7×11	(7) 4×8	(8) 2×4	(9) 2×10	(10) 4×3
(11) 11×8	(12) 12×8	(13) 4×2	(14) 12×5	(15) 3×7
(16) 3×2	(17) 6×2	(18) 7×10	(19) 4×5	(20) 9×2
(21) 11×9	(22) 5×8	(23) 0×9	(24) 11×11	(25) 7×7
(26) 2×12	(27) 9×6	(28) 12×7	(29) 1×2	(30) 9×9
(31) 6×0	(32) 12×9	(33) 2×8	(34) 12×2	(35) 11×4
(36) 1×3	(37) 4×11	(38) 1×4	(39) 0×2	(40) 11×2

92

Multiply by 0-12

Time: ________

Score: _____/ 40

(1) 4 × 10	(2) 11 × 7	(3) 3 × 9	(4) 3 × 0	(5) 1 × 0
(6) 11 × 6	(7) 7 × 1	(8) 7 × 2	(9) 8 × 11	(10) 12 × 8
(11) 10 × 5	(12) 1 × 9	(13) 7 × 0	(14) 7 × 5	(15) 12 × 11
(16) 5 × 2	(17) 10 × 4	(18) 10 × 3	(19) 0 × 4	(20) 5 × 7
(21) 3 × 1	(22) 10 × 8	(23) 8 × 0	(24) 1 × 6	(25) 9 × 8
(26) 2 × 6	(27) 5 × 10	(28) 0 × 9	(29) 9 × 12	(30) 9 × 7
(31) 8 × 10	(32) 4 × 7	(33) 1 × 8	(34) 3 × 7	(35) 5 × 9
(36) 6 × 3	(37) 11 × 0	(38) 5 × 11	(39) 0 × 7	(40) 5 × 6

93

Multiply by 0-12

Time: ________

Score: _____/ 40

(1) 1×9	(2) 11×3	(3) 8×5	(4) 4×9	(5) 11×0
(6) 2×10	(7) 3×5	(8) 12×10	(9) 4×4	(10) 12×1
(11) 9×6	(12) 1×5	(13) 8×6	(14) 10×12	(15) 10×11
(16) 2×12	(17) 12×4	(18) 1×12	(19) 9×8	(20) 2×2
(21) 0×0	(22) 0×4	(23) 3×4	(24) 3×9	(25) 10×3
(26) 8×8	(27) 11×1	(28) 0×11	(29) 7×1	(30) 3×7
(31) 1×3	(32) 6×12	(33) 4×10	(34) 1×11	(35) 4×12
(36) 9×3	(37) 4×0	(38) 11×4	(39) 5×3	(40) 7×0

94

Multiply by 0-12

Time: ________

Score: _____/ 40

(1) 12×12	(2) 1×5	(3) 3×11	(4) 2×0	(5) 6×0
(6) 11×8	(7) 3×12	(8) 9×0	(9) 1×1	(10) 2×6
(11) 12×7	(12) 3×0	(13) 0×10	(14) 9×12	(15) 6×5
(16) 6×9	(17) 7×10	(18) 10×11	(19) 4×0	(20) 12×10
(21) 10×0	(22) 9×6	(23) 9×2	(24) 5×11	(25) 9×5
(26) 3×2	(27) 5×6	(28) 0×10	(29) 0×6	(30) 7×12
(31) 9×8	(32) 1×9	(33) 4×5	(34) 12×5	(35) 11×12
(36) 6×10	(37) 8×7	(38) 9×11	(39) 4×6	(40) 12×0

95

Multiply by 0-12

Time: __________

Score: _____/ 40

(1) 10×0	(2) 4×6	(3) 5×3	(4) 9×12	(5) 0×5
(6) 10×9	(7) 8×5	(8) 10×10	(9) 0×0	(10) 3×7
(11) 5×9	(12) 10×11	(13) 11×9	(14) 1×10	(15) 1×0
(16) 9×1	(17) 2×12	(18) 7×9	(19) 6×3	(20) 5×5
(21) 0×8	(22) 6×6	(23) 6×8	(24) 7×7	(25) 11×3
(26) 7×1	(27) 10×1	(28) 1×12	(29) 8×8	(30) 2×0
(31) 8×12	(32) 9×7	(33) 12×4	(34) 1×4	(35) 7×8
(36) 11×1	(37) 7×4	(38) 9×8	(39) 7×10	(40) 1×10

96

Multiply by 0-12

Time: ________

Score: _____/ 40

(1) 8×12	(2) 3×8	(3) 10×12	(4) 7×6	(5) 5×1
(6) 5×7	(7) 11×7	(8) 9×9	(9) 10×2	(10) 4×4
(11) 7×11	(12) 5×6	(13) 11×6	(14) 4×9	(15) 8×10
(16) 0×6	(17) 9×4	(18) 3×0	(19) 2×8	(20) 2×3
(21) 4×3	(22) 12×11	(23) 5×11	(24) 10×8	(25) 2×4
(26) 8×3	(27) 7×4	(28) 1×2	(29) 11×0	(30) 9×10
(31) 1×4	(32) 12×3	(33) 2×11	(34) 0×9	(35) 10×3
(36) 11×2	(37) 0×1	(38) 5×12	(39) 2×10	(40) 2×5

97

Multiply by 0-12

Time: ________

Score: _____/ 40

(1) 3 × 6	(2) 12 × 0	(3) 0 × 12	(4) 8 × 4	(5) 5 × 0
(6) 6 × 1	(7) 10 × 6	(8) 6 × 12	(9) 5 × 8	(10) 4 × 8
(11) 6 × 3	(12) 1 × 3	(13) 4 × 12	(14) 7 × 0	(15) 6 × 8
(16) 8 × 4	(17) 3 × 9	(18) 11 × 6	(19) 7 × 5	(20) 11 × 8
(21) 8 × 0	(22) 11 × 11	(23) 9 × 2	(24) 3 × 4	(25) 7 × 2
(26) 6 × 2	(27) 4 × 5	(28) 4 × 11	(29) 10 × 4	(30) 4 × 10
(31) 4 × 3	(32) 4 × 0	(33) 3 × 4	(34) 3 × 3	(35) 12 × 4
(36) 5 × 4	(37) 0 × 4	(38) 4 × 8	(39) 9 × 11	(40) 9 × 9

98
Multiply by 0-12

Time: __________

Score: _____/ 40

① 4 × 2	② 3 × 0	③ 8 × 1	④ 10 × 6	⑤ 2 × 4
⑥ 10 × 9	⑦ 1 × 5	⑧ 11 × 1	⑨ 10 × 4	⑩ 1 × 8
⑪ 9 × 0	⑫ 9 × 4	⑬ 11 × 0	⑭ 9 × 8	⑮ 7 × 1
⑯ 10 × 7	⑰ 4 × 10	⑱ 12 × 3	⑲ 7 × 0	⑳ 10 × 10
(21) 1 × 10	(22) 11 × 8	(23) 2 × 9	(24) 10 × 12	(25) 9 × 1
(26) 4 × 2	(27) 7 × 12	(28) 7 × 5	(29) 7 × 4	(30) 11 × 2
(31) 11 × 10	(32) 8 × 0	(33) 5 × 1	(34) 0 × 0	(35) 10 × 1
(36) 0 × 11	(37) 5 × 6	(38) 11 × 12	(39) 6 × 3	(40) 10 × 4

99

Multiply by 0-12

Time: ________

Score: _____/ 40

(1) 12 × 9	(2) 2 × 2	(3) 4 × 12	(4) 12 × 8	(5) 4 × 4
(6) 12 × 5	(7) 8 × 7	(8) 3 × 2	(9) 0 × 10	(10) 1 × 12
(11) 2 × 8	(12) 12 × 12	(13) 8 × 12	(14) 4 × 4	(15) 0 × 4
(16) 5 × 8	(17) 6 × 6	(18) 11 × 4	(19) 0 × 1	(20) 4 × 3
(21) 9 × 3	(22) 11 × 6	(23) 8 × 6	(24) 0 × 9	(25) 10 × 12
(26) 7 × 5	(27) 10 × 9	(28) 10 × 0	(29) 3 × 4	(30) 5 × 12
(31) 11 × 12	(32) 2 × 6	(33) 0 × 3	(34) 8 × 2	(35) 3 × 9
(36) 5 × 5	(37) 11 × 9	(38) 11 × 11	(39) 8 × 5	(40) 7 × 1

100
Multiply by 0-12

Time: ________

Score: _____/ 40

(1) 12×0	(2) 11×5	(3) 8×10	(4) 10×3	(5) 11×6
(6) 5×0	(7) 3×1	(8) 11×4	(9) 5×11	(10) 6×2
(11) 4×6	(12) 9×1	(13) 0×5	(14) 12×7	(15) 1×11
(16) 9×12	(17) 7×4	(18) 7×7	(19) 4×1	(20) 4×0
(21) 5×4	(22) 8×0	(23) 0×12	(24) 9×4	(25) 4×12
(26) 8×2	(27) 11×7	(28) 7×9	(29) 0×2	(30) 12×4
(31) 0×1	(32) 1×4	(33) 12×0	(34) 9×2	(35) 2×0
(36) 9×7	(37) 5×3	(38) 8×8	(39) 7×11	(40) 6×12

Answer Key # 1

1. 0	2. 0	3. 0	4. 0	5. 9
6. 0	7. 0	8. 1	9. 7	10. 0
11. 0	12. 0	13. 0	14. 5	15. 6
16. 0	17. 3	18. 8	19. 0	20. 0
21. 5	22. 0	23. 0	24. 0	25. 9
26. 0	27. 6	28. 0	29. 0	30. 8
31. 0	32. 7	33. 0	34. 0	35. 3
36. 0	37. 0	38. 0	39. 0	40. 1

Answer Key # 2

1. 0	2. 3	3. 0	4. 0	5. 5
6. 7	7. 4	8. 0	9. 0	10. 6
11. 0	12. 0	13. 1	14. 4	15. 0
16. 0	17. 0	18. 9	19. 8	20. 0
21. 0	22. 5	23. 7	24. 1	25. 9
26. 0	27. 6	28. 0	29. 0	30. 4
31. 0	32. 0	33. 0	34. 4	35. 8
36. 0	37. 0	38. 3	39. 0	40. 0

Answer Key # 3

1. 6	2. 0	3. 8	4. 5	5. 7
6. 0	7. 0	8. 0	9. 0	10. 0
11. 9	12. 4	13. 0	14. 0	15. 0
16. 4	17. 3	18. 1	19. 0	20. 0
21. 0	22. 0	23. 4	24. 0	25. 3
26. 0	27. 1	28. 0	29. 0	30. 0
31. 7	32. 0	33. 0	34. 8	35. 9
36. 4	37. 6	38. 5	39. 0	40. 0

Answer Key # 4

1. 1	2. 5	3. 3	4. 2	5. 7
6. 4	7. 0	8. 0	9. 0	10. 0
11. 6	12. 0	13. 0	14. 4	15. 2
16. 0	17. 8	18. 9	19. 0	20. 0
21. 4	22. 3	23. 0	24. 2	25. 0
26. 5	27. 8	28. 0	29. 7	30. 4
31. 0	32. 9	33. 0	34. 2	35. 0
36. 0	37. 0	38. 0	39. 1	40. 6

Answer Key # 5

1. 0	2. 8	3. 10	4. 2	5. 18
6. 16	7. 6	8. 14	9. 4	10. 12
11. 10	12. 14	13. 6	14. 12	15. 8
16. 0	17. 4	18. 18	19. 16	20. 2
21. 12	22. 8	23. 18	24. 16	25. 0
26. 2	27. 4	28. 6	29. 14	30. 10
31. 16	32. 12	33. 8	34. 0	35. 14
36. 18	37. 4	38. 10	39. 6	40. 2

Answer Key # 6

1. 8	2. 4	3. 18	4. 0	5. 14
6. 12	7. 16	8. 6	9. 10	10. 2
11. 12	12. 18	13. 6	14. 10	15. 16
16. 0	17. 4	18. 14	19. 2	20. 8
21. 6	22. 10	23. 2	24. 8	25. 16
26. 14	27. 12	28. 18	29. 0	30. 4
31. 0	32. 2	33. 14	34. 10	35. 18
36. 12	37. 16	38. 4	39. 6	40. 8

Answer Key # 7

1. 2	2. 16	3. 12	4. 8	5. 4
6. 14	7. 18	8. 0	9. 10	10. 6
11. 14	12. 8	13. 12	14. 16	15. 6
16. 2	17. 10	18. 0	19. 18	20. 4
21. 6	22. 0	23. 2	24. 8	25. 12
26. 10	27. 4	28. 16	29. 14	30. 18
31. 14	32. 6	33. 16	34. 2	35. 12
36. 4	37. 8	38. 18	39. 10	40. 0

Answer Key # 8

1. 18	2. 6	3. 0	4. 12	5. 2
6. 14	7. 8	8. 10	9. 16	10. 4
11. 18	12. 6	13. 0	14. 4	15. 10
16. 14	17. 12	18. 16	19. 8	20. 2
21. 16	22. 4	23. 12	24. 8	25. 6
26. 0	27. 10	28. 2	29. 18	30. 14
31. 18	32. 8	33. 6	34. 0	35. 16
36. 12	37. 10	38. 14	39. 2	40. 4

Answer Key # 9

1. 18	2. 12	3. 15	4. 6	5. 0
6. 21	7. 3	8. 9	9. 24	10. 27
11. 21	12. 12	13. 3	14. 15	15. 27
16. 0	17. 18	18. 6	19. 24	20. 9
21. 21	22. 3	23. 0	24. 6	25. 18
26. 15	27. 9	28. 27	29. 24	30. 12
31. 27	32. 18	33. 24	34. 12	35. 3
36. 9	37. 21	38. 6	39. 15	40. 0

Answer Key # 10

1. 12	2. 9	3. 21	4. 6	5. 3
6. 27	7. 24	8. 15	9. 18	10. 0
11. 18	12. 27	13. 15	14. 0	15. 21
16. 12	17. 6	18. 24	19. 3	20. 9
21. 6	22. 9	23. 24	24. 12	25. 15
26. 0	27. 18	28. 27	29. 3	30. 21
31. 12	32. 21	33. 15	34. 27	35. 0
36. 18	37. 24	38. 3	39. 9	40. 6

Answer Key # 11

1. 0	2. 15	3. 9	4. 24	5. 21
6. 27	7. 12	8. 6	9. 18	10. 3
11. 18	12. 12	13. 15	14. 6	15. 3
16. 27	17. 0	18. 21	19. 9	20. 24
21. 18	22. 6	23. 27	24. 3	25. 12
26. 0	27. 24	28. 15	29. 9	30. 21
31. 24	32. 0	33. 3	34. 12	35. 21
36. 6	37. 18	38. 9	39. 15	40. 27

Answer Key # 12

1. 21	2. 27	3. 9	4. 24	5. 12
6. 15	7. 3	8. 6	9. 0	10. 18
11. 15	12. 21	13. 18	14. 6	15. 24
16. 3	17. 27	18. 9	19. 0	20. 12
21. 21	22. 12	23. 27	24. 24	25. 0
26. 3	27. 15	28. 18	29. 6	30. 9
31. 21	32. 18	33. 0	34. 6	35. 3
36. 15	37. 24	38. 12	39. 9	40. 27

Answer Key # 13

1. 3	2. 2	3. 0	4. 0	5. 0
6. 9	7. 1	8. 27	9. 0	10. 0
11. 4	12. 10	13. 12	14. 0	15. 6
16. 18	17. 7	18. 8	19. 14	20. 21
21. 6	22. 12	23. 0	24. 12	25. 24
26. 3	27. 2	28. 8	29. 0	30. 18
31. 15	32. 18	33. 0	34. 0	35. 5
36. 2	37. 4	38. 0	39. 6	40. 9

Answer Key # 14

1. 0	2. 0	3. 21	4. 4	5. 0
6. 9	7. 12	8. 0	9. 3	10. 0
11. 4	12. 14	13. 12	14. 0	15. 27
16. 0	17. 24	18. 5	19. 0	20. 18
21. 1	22. 0	23. 6	24. 0	25. 9
26. 7	27. 3	28. 12	29. 6	30. 4
31. 0	32. 18	33. 8	34. 10	35. 6
36. 4	37. 15	38. 18	39. 8	40. 0

Answer Key # 15

1. 9	2. 6	3. 10	4. 4	5. 0
6. 21	7. 7	8. 6	9. 0	10. 0
11. 18	12. 14	13. 8	14. 3	15. 0
16. 4	17. 15	18. 0	19. 6	20. 0
21. 0	22. 6	23. 5	24. 0	25. 6
26. 18	27. 0	28. 0	29. 1	30. 24
31. 8	32. 0	33. 18	34. 12	35. 0
36. 3	37. 9	38. 27	39. 0	40. 0

Answer Key # 16

1. 4	2. 7	3. 0	4. 5	5. 0
6. 21	7. 3	8. 9	9. 4	10. 2
11. 0	12. 12	13. 6	14. 0	15. 0
16. 0	17. 12	18. 16	19. 12	20. 0
21. 2	22. 24	23. 0	24. 6	25. 0
26. 0	27. 1	28. 15	29. 4	30. 3
31. 18	32. 0	33. 4	34. 16	35. 9
36. 10	37. 0	38. 27	39. 14	40. 2

Answer Key # 17

1. 0	2. 4	3. 0	4. 3	5. 0
6. 0	7. 15	8. 7	9. 0	10. 12
11. 3	12. 24	13. 12	14. 18	15. 0
16. 9	17. 0	18. 4	19. 2	20. 21
21. 10	22. 6	23. 12	24. 18	25. 0
26. 9	27. 18	28. 0	29. 0	30. 6
31. 2	32. 0	33. 0	34. 5	35. 1
36. 14	37. 0	38. 0	39. 2	40. 27

Answer Key # 18

1. 9	2. 16	3. 0	4. 21	5. 24
6. 10	7. 0	8. 0	9. 8	10. 16
11. 0	12. 1	13. 0	14. 4	15. 0
16. 18	17. 8	18. 27	19. 0	20. 15
21. 0	22. 3	23. 12	24. 0	25. 0
26. 6	27. 5	28. 0	29. 4	30. 0
31. 9	32. 7	33. 14	34. 0	35. 8
36. 0	37. 6	38. 3	39. 6	40. 8

Answer Key # 19

1. 0	2. 7	3. 8	4. 15	5. 0
6. 4	7. 0	8. 21	9. 0	10. 10
11. 0	12. 4	13. 12	14. 1	15. 9
16. 0	17. 0	18. 0	19. 0	20. 14
21. 5	22. 6	23. 0	24. 8	25. 18
26. 27	27. 0	28. 9	29. 0	30. 12
31. 6	32. 12	33. 0	34. 18	35. 3
36. 3	37. 0	38. 18	39. 24	40. 6

Answer Key # 20

1. 2	2. 0	3. 0	4. 6	5. 27
6. 0	7. 18	8. 10	9. 2	10. 0
11. 0	12. 24	13. 9	14. 21	15. 0
16. 12	17. 12	18. 0	19. 5	20. 0
21. 16	22. 9	23. 7	24. 15	25. 3
26. 0	27. 2	28. 0	29. 0	30. 0
31. 0	32. 0	33. 12	34. 1	35. 0
36. 3	37. 16	38. 0	39. 6	40. 14

Answer Key # 21

1. 28	2. 20	3. 16	4. 12	5. 32
6. 4	7. 0	8. 8	9. 36	10. 24
11. 12	12. 20	13. 0	14. 4	15. 32
16. 16	17. 8	18. 24	19. 28	20. 36
21. 12	22. 36	23. 8	24. 32	25. 4
26. 16	27. 28	28. 0	29. 20	30. 24
31. 36	32. 4	33. 32	34. 20	35. 12
36. 28	37. 8	38. 0	39. 24	40. 16

Answer Key # 22

1. 32	2. 16	3. 36	4. 20	5. 12
6. 28	7. 24	8. 4	9. 8	10. 0
11. 8	12. 0	13. 28	14. 12	15. 24
16. 36	17. 32	18. 20	19. 16	20. 4
21. 8	22. 0	23. 24	24. 16	25. 12
26. 28	27. 36	28. 32	29. 4	30. 20
31. 28	32. 20	33. 12	34. 36	35. 8
36. 0	37. 4	38. 16	39. 24	40. 32

Answer Key # 23

1. 24	2. 32	3. 12	4. 4	5. 8
6. 0	7. 20	8. 36	9. 16	10. 28
11. 0	12. 8	13. 32	14. 16	15. 24
16. 4	17. 28	18. 36	19. 20	20. 12
21. 24	22. 12	23. 28	24. 32	25. 8
26. 4	27. 36	28. 20	29. 0	30. 16
31. 24	32. 16	33. 0	34. 8	35. 4
36. 12	37. 36	38. 28	39. 20	40. 32

Answer Key # 24

1. 4	2. 24	3. 28	4. 16	5. 32
6. 12	7. 36	8. 8	9. 20	10. 0
11. 36	12. 28	13. 32	14. 8	15. 4
16. 0	17. 24	18. 12	19. 16	20. 20
21. 32	22. 0	23. 20	24. 16	25. 28
26. 4	27. 8	28. 24	29. 36	30. 12
31. 4	32. 12	33. 20	34. 24	35. 36
36. 0	37. 28	38. 16	39. 8	40. 32

Answer Key # 25

1. 45	2. 25	3. 5	4. 0	5. 10
6. 35	7. 30	8. 15	9. 20	10. 40
11. 45	12. 0	13. 40	14. 35	15. 25
16. 30	17. 5	18. 10	19. 15	20. 20
21. 0	22. 40	23. 10	24. 15	25. 30
26. 5	27. 20	28. 25	29. 35	30. 45
31. 40	32. 30	33. 45	34. 10	35. 5
36. 20	37. 25	38. 35	39. 15	40. 0

Answer Key # 26

1. 35	2. 0	3. 5	4. 40	5. 45
6. 15	7. 10	8. 25	9. 30	10. 20
11. 45	12. 5	13. 25	14. 35	15. 40
16. 30	17. 20	18. 15	19. 10	20. 0
21. 35	22. 30	23. 0	24. 15	25. 10
26. 20	27. 45	28. 40	29. 5	30. 25
31. 35	32. 25	33. 10	34. 40	35. 30
36. 5	37. 15	38. 20	39. 0	40. 45

Answer Key # 27

1. 5	2. 30	3. 35	4. 25	5. 20
6. 15	7. 0	8. 40	9. 45	10. 10
11. 25	12. 0	13. 10	14. 45	15. 35
16. 40	17. 20	18. 30	19. 15	20. 5
21. 20	22. 35	23. 10	24. 45	25. 30
26. 5	27. 15	28. 40	29. 0	30. 25
31. 45	32. 40	33. 10	34. 25	35. 20
36. 35	37. 5	38. 15	39. 0	40. 30

Answer Key # 28

1. 0	2. 40	3. 45	4. 35	5. 5
6. 25	7. 20	8. 10	9. 30	10. 15
11. 0	12. 15	13. 25	14. 10	15. 35
16. 45	17. 20	18. 40	19. 5	20. 30
21. 20	22. 35	23. 5	24. 0	25. 45
26. 10	27. 25	28. 40	29. 30	30. 15
31. 0	32. 15	33. 10	34. 25	35. 45
36. 5	37. 20	38. 35	39. 40	40. 30

Answer Key # 29

1. 12	2. 54	3. 6	4. 0	5. 24
6. 30	7. 42	8. 36	9. 18	10. 48
11. 6	12. 24	13. 12	14. 30	15. 48
16. 36	17. 54	18. 42	19. 18	20. 0
21. 36	22. 30	23. 6	24. 54	25. 0
26. 48	27. 12	28. 42	29. 18	30. 24
31. 30	32. 24	33. 48	34. 6	35. 54
36. 12	37. 42	38. 0	39. 18	40. 36

Answer Key # 30

1. 30	2. 42	3. 0	4. 18	5. 6
6. 54	7. 36	8. 24	9. 12	10. 48
11. 6	12. 42	13. 24	14. 30	15. 36
16. 12	17. 0	18. 18	19. 54	20. 48
21. 12	22. 24	23. 30	24. 48	25. 54
26. 42	27. 36	28. 6	29. 18	30. 0
31. 42	32. 54	33. 12	34. 24	35. 6
36. 30	37. 18	38. 0	39. 48	40. 36

Answer Key # 31

1. 30	2. 36	3. 0	4. 6	5. 54
6. 24	7. 12	8. 42	9. 48	10. 18
11. 48	12. 18	13. 54	14. 30	15. 12
16. 36	17. 24	18. 42	19. 6	20. 0
21. 42	22. 18	23. 0	24. 54	25. 48
26. 30	27. 12	28. 6	29. 24	30. 36
31. 42	32. 6	33. 24	34. 36	35. 48
36. 12	37. 54	38. 30	39. 18	40. 0

Answer Key # 32

1. 12	2. 54	3. 30	4. 0	5. 24
6. 48	7. 6	8. 42	9. 36	10. 18
11. 42	12. 12	13. 6	14. 30	15. 24
16. 18	17. 36	18. 48	19. 54	20. 0
21. 24	22. 18	23. 36	24. 12	25. 30
26. 42	27. 48	28. 54	29. 0	30. 6
31. 12	32. 54	33. 0	34. 30	35. 24
36. 36	37. 6	38. 42	39. 18	40. 48

Answer Key # 33

1. 0	2. 10	3. 12	4. 16	5. 4
6. 42	7. 48	8. 40	9. 30	10. 24
11. 36	12. 32	13. 20	14. 24	15. 28
16. 35	17. 18	18. 30	19. 20	20. 15
21. 8	22. 45	23. 12	24. 0	25. 0
26. 5	27. 25	28. 54	29. 36	30. 8
31. 30	32. 35	33. 0	34. 20	35. 40
36. 8	37. 25	38. 30	39. 32	40. 42

Answer Key # 34

1. 10	2. 36	3. 28	4. 5	5. 15
6. 18	7. 20	8. 0	9. 0	10. 36
11. 54	12. 12	13. 4	14. 8	15. 24
16. 12	17. 16	18. 48	19. 24	20. 45
21. 48	22. 16	23. 10	24. 36	25. 0
26. 24	27. 20	28. 40	29. 28	30. 12
31. 45	32. 20	33. 4	34. 30	35. 35
36. 12	37. 42	38. 15	39. 30	40. 0

Answer Key # 35

1. 5	2. 24	3. 54	4. 6	5. 6
6. 0	7. 32	8. 25	9. 36	10. 18
11. 32	12. 0	13. 24	14. 30	15. 15
16. 16	17. 30	18. 28	19. 10	20. 48
21. 36	22. 6	23. 25	24. 54	25. 0
26. 45	27. 12	28. 5	29. 24	30. 40
31. 0	32. 6	33. 36	34. 35	35. 20
36. 42	37. 18	38. 4	39. 20	40. 12

Answer Key # 36

1. 24	2. 40	3. 18	4. 54	5. 32
6. 5	7. 0	8. 36	9. 0	10. 48
11. 42	12. 30	13. 25	14. 45	15. 20
16. 24	17. 28	18. 35	19. 15	20. 0
21. 10	22. 20	23. 12	24. 8	25. 16
26. 4	27. 30	28. 8	29. 12	30. 36
31. 54	32. 4	33. 24	34. 30	35. 18
36. 0	37. 28	38. 24	39. 30	40. 35

Answer Key # 37

1. 36	2. 15	3. 12	4. 20	5. 20
6. 36	7. 8	8. 10	9. 0	10. 16
11. 0	12. 45	13. 12	14. 8	15. 5
16. 48	17. 40	18. 42	19. 32	20. 25
21. 30	22. 32	23. 36	24. 42	25. 36
26. 12	27. 0	28. 20	29. 35	30. 4
31. 25	32. 8	33. 54	34. 24	35. 48
36. 8	37. 0	38. 12	39. 24	40. 10

Answer Key # 38

1. 16	2. 5	3. 30	4. 0	5. 45
6. 20	7. 40	8. 18	9. 28	10. 15
11. 18	12. 24	13. 24	14. 0	15. 5
16. 36	17. 30	18. 32	19. 8	20. 12
21. 25	22. 4	23. 36	24. 45	25. 48
26. 12	27. 54	28. 20	29. 8	30. 10
31. 15	32. 20	33. 35	34. 42	35. 40
36. 0	37. 28	38. 30	39. 0	40. 16

Answer Key # 39

1. 28	2. 25	3. 6	4. 20	5. 15
6. 24	7. 20	8. 5	9. 35	10. 30
11. 45	12. 12	13. 36	14. 10	15. 18
16. 42	17. 36	18. 48	19. 54	20. 12
21. 32	22. 24	23. 40	24. 0	25. 0
26. 30	27. 4	28. 6	29. 0	30. 16
31. 45	32. 20	33. 15	34. 0	35. 20
36. 36	37. 54	38. 48	39. 10	40. 0

Answer Key # 40

1. 32	2. 40	3. 42	4. 25	5. 36
6. 24	7. 30	8. 5	9. 12	10. 12
11. 35	12. 8	13. 8	14. 30	15. 4
16. 18	17. 16	18. 28	19. 0	20. 24
21. 32	22. 36	23. 20	24. 18	25. 35
26. 0	27. 12	28. 30	29. 40	30. 24
31. 0	32. 36	33. 25	34. 28	35. 8
36. 10	37. 42	38. 12	39. 24	40. 0

Answer Key # 41

1. 21	2. 49	3. 7	4. 28	5. 56
6. 35	7. 63	8. 42	9. 14	10. 0
11. 7	12. 49	13. 63	14. 0	15. 14
16. 56	17. 35	18. 28	19. 21	20. 42
21. 28	22. 14	23. 21	24. 35	25. 7
26. 49	27. 56	28. 0	29. 42	30. 63
31. 0	32. 14	33. 28	34. 35	35. 49
36. 63	37. 7	38. 42	39. 56	40. 21

Answer Key # 42

1. 0	2. 35	3. 42	4. 28	5. 49
6. 21	7. 56	8. 14	9. 63	10. 7
11. 0	12. 42	13. 35	14. 7	15. 21
16. 63	17. 56	18. 14	19. 28	20. 49
21. 63	22. 28	23. 56	24. 49	25. 7
26. 14	27. 0	28. 35	29. 42	30. 21
31. 42	32. 21	33. 35	34. 56	35. 49
36. 63	37. 0	38. 28	39. 14	40. 7

Answer Key # 43

1. 28	2. 35	3. 7	4. 42	5. 0
6. 14	7. 21	8. 49	9. 56	10. 63
11. 42	12. 49	13. 7	14. 0	15. 14
16. 28	17. 56	18. 35	19. 21	20. 63
21. 28	22. 63	23. 56	24. 0	25. 42
26. 7	27. 35	28. 49	29. 21	30. 14
31. 21	32. 0	33. 56	34. 49	35. 63
36. 35	37. 7	38. 28	39. 42	40. 14

Answer Key # 44

1. 7	2. 49	3. 21	4. 42	5. 28
6. 0	7. 63	8. 35	9. 56	10. 14
11. 0	12. 49	13. 35	14. 14	15. 21
16. 28	17. 56	18. 42	19. 7	20. 63
21. 35	22. 7	23. 21	24. 14	25. 0
26. 49	27. 28	28. 63	29. 56	30. 42
31. 63	32. 35	33. 56	34. 49	35. 21
36. 14	37. 7	38. 28	39. 0	40. 42

Answer Key # 45

1. 40	2. 32	3. 24	4. 48	5. 8
6. 72	7. 16	8. 64	9. 0	10. 56
11. 24	12. 56	13. 64	14. 40	15. 32
16. 16	17. 72	18. 0	19. 8	20. 48
21. 64	22. 48	23. 32	24. 40	25. 56
26. 72	27. 0	28. 8	29. 16	30. 24
31. 40	32. 72	33. 32	34. 64	35. 16
36. 48	37. 24	38. 0	39. 8	40. 56

Answer Key # 46

1. 72	2. 16	3. 8	4. 32	5. 64
6. 0	7. 40	8. 24	9. 56	10. 48
11. 72	12. 24	13. 8	14. 56	15. 32
16. 48	17. 16	18. 0	19. 64	20. 40
21. 64	22. 48	23. 24	24. 40	25. 8
26. 56	27. 72	28. 16	29. 32	30. 0
31. 16	32. 56	33. 0	34. 8	35. 40
36. 48	37. 72	38. 64	39. 32	40. 24

Answer Key # 47

1. 72	2. 24	3. 8	4. 32	5. 40
6. 16	7. 48	8. 64	9. 0	10. 56
11. 40	12. 0	13. 16	14. 64	15. 24
16. 8	17. 32	18. 56	19. 48	20. 72
21. 40	22. 72	23. 48	24. 64	25. 32
26. 16	27. 0	28. 8	29. 24	30. 56
31. 32	32. 56	33. 64	34. 24	35. 48
36. 16	37. 0	38. 8	39. 40	40. 72

Answer Key # 48

1. 64	2. 8	3. 16	4. 0	5. 48
6. 32	7. 40	8. 24	9. 72	10. 56
11. 0	12. 8	13. 72	14. 16	15. 56
16. 48	17. 40	18. 64	19. 24	20. 32
21. 40	22. 32	23. 56	24. 24	25. 0
26. 64	27. 48	28. 8	29. 72	30. 16
31. 0	32. 32	33. 56	34. 16	35. 48
36. 40	37. 8	38. 64	39. 24	40. 72

Answer Key # 49

1. 81	2. 36	3. 18	4. 54	5. 0
6. 72	7. 27	8. 45	9. 9	10. 63
11. 18	12. 54	13. 27	14. 81	15. 63
16. 0	17. 36	18. 9	19. 72	20. 45
21. 36	22. 45	23. 9	24. 72	25. 27
26. 63	27. 0	28. 81	29. 54	30. 18
31. 81	32. 18	33. 72	34. 63	35. 9
36. 0	37. 54	38. 45	39. 27	40. 36

Answer Key # 50

1. 27	2. 9	3. 0	4. 63	5. 54
6. 81	7. 45	8. 72	9. 18	10. 36
11. 63	12. 36	13. 0	14. 45	15. 81
16. 18	17. 9	18. 72	19. 27	20. 54
21. 9	22. 81	23. 45	24. 63	25. 54
26. 27	27. 72	28. 0	29. 36	30. 18
31. 9	32. 45	33. 36	34. 63	35. 18
36. 27	37. 72	38. 81	39. 0	40. 54

Answer Key # 51

1. 36	2. 72	3. 0	4. 81	5. 63
6. 18	7. 9	8. 54	9. 45	10. 27
11. 72	12. 18	13. 81	14. 0	15. 45
16. 63	17. 9	18. 54	19. 27	20. 36
21. 18	22. 45	23. 9	24. 63	25. 81
26. 54	27. 36	28. 72	29. 0	30. 27
31. 54	32. 72	33. 18	34. 36	35. 27
36. 81	37. 63	38. 45	39. 9	40. 0

Answer Key # 52

1. 63	2. 36	3. 9	4. 27	5. 54
6. 45	7. 18	8. 0	9. 81	10. 72
11. 9	12. 72	13. 63	14. 0	15. 36
16. 81	17. 27	18. 45	19. 18	20. 54
21. 63	22. 54	23. 9	24. 27	25. 45
26. 18	27. 0	28. 72	29. 81	30. 36
31. 9	32. 54	33. 72	34. 27	35. 63
36. 45	37. 36	38. 0	39. 18	40. 81

Answer Key # 53

1. 63	2. 36	3. 90	4. 81	5. 54
6. 21	7. 63	8. 9	9. 35	10. 56
11. 28	12. 16	13. 0	14. 72	15. 0
16. 8	17. 48	18. 70	19. 80	20. 18
21. 49	22. 56	23. 0	24. 32	25. 27
26. 14	27. 24	28. 45	29. 42	30. 40
31. 72	32. 64	33. 7	34. 8	35. 0
36. 21	37. 64	38. 27	39. 80	40. 7

Answer Key # 54

1. 36	2. 40	3. 42	4. 56	5. 32
6. 90	7. 48	8. 9	9. 45	10. 0
11. 35	12. 24	13. 72	14. 14	15. 72
16. 18	17. 28	18. 63	19. 63	20. 0
21. 16	22. 54	23. 49	24. 81	25. 56
26. 70	27. 0	28. 49	29. 8	30. 0
31. 63	32. 40	33. 54	34. 7	35. 56
36. 24	37. 9	38. 48	39. 80	40. 16

Answer Key # 55

1. 36	2. 35	3. 45	4. 32	5. 72
6. 81	7. 28	8. 42	9. 14	10. 90
11. 64	12. 56	13. 63	14. 0	15. 72
16. 27	17. 18	18. 21	19. 70	20. 81
21. 63	22. 63	23. 40	24. 56	25. 18
26. 72	27. 27	28. 45	29. 9	30. 8
31. 64	32. 0	33. 70	34. 21	35. 72
36. 16	37. 56	38. 80	39. 0	40. 54

Answer Key # 56

1. 35	2. 90	3. 42	4. 48	5. 24
6. 32	7. 0	8. 49	9. 14	10. 7
11. 36	12. 28	13. 16	14. 8	15. 42
16. 32	17. 36	18. 63	19. 7	20. 9
21. 90	22. 18	23. 21	24. 81	25. 35
26. 54	27. 0	28. 49	29. 56	30. 72
31. 48	32. 24	33. 0	34. 0	35. 27
36. 70	37. 64	38. 72	39. 14	40. 63

Answer Key # 57

1. 56	2. 40	3. 80	4. 45	5. 28
6. 18	7. 27	8. 40	9. 0	10. 35
11. 36	12. 64	13. 45	14. 42	15. 63
16. 8	17. 80	18. 32	19. 90	20. 56
21. 7	22. 24	23. 81	24. 9	25. 70
26. 56	27. 72	28. 72	29. 21	30. 48
31. 16	32. 63	33. 49	34. 0	35. 28
36. 0	37. 54	38. 14	39. 80	40. 45

Answer Key # 58

1. 24	2. 18	3. 56	4. 49	5. 16
6. 56	7. 42	8. 40	9. 54	10. 21
11. 63	12. 81	13. 70	14. 35	15. 0
16. 0	17. 72	18. 28	19. 7	20. 9
21. 64	22. 27	23. 36	24. 72	25. 48
26. 0	27. 90	28. 63	29. 14	30. 8
31. 32	32. 72	33. 80	34. 0	35. 7
36. 90	37. 27	38. 8	39. 36	40. 63

Answer Key # 59

1. 21	2. 56	3. 18	4. 54	5. 42
6. 48	7. 9	8. 56	9. 16	10. 64
11. 0	12. 0	13. 49	14. 28	15. 81
16. 70	17. 24	18. 14	19. 45	20. 35
21. 32	22. 40	23. 72	24. 63	25. 40
26. 27	27. 81	28. 9	29. 54	30. 63
31. 0	32. 72	33. 42	34. 80	35. 28
36. 45	37. 49	38. 32	39. 56	40. 72

Answer Key # 60

1. 56	2. 35	3. 63	4. 70	5. 18
6. 16	7. 64	8. 36	9. 90	10. 0
11. 7	12. 8	13. 21	14. 14	15. 0
16. 48	17. 24	18. 64	19. 63	20. 81
21. 18	22. 40	23. 35	24. 0	25. 7
26. 24	27. 90	28. 63	29. 14	30. 45
31. 32	32. 28	33. 48	34. 9	35. 16
36. 70	37. 56	38. 56	39. 54	40. 27

Answer Key # 61

1. 40	2. 80	3. 90	4. 0	5. 30
6. 10	7. 100	8. 60	9. 50	10. 20
11. 70	12. 50	13. 20	14. 90	15. 40
16. 0	17. 30	18. 80	19. 70	20. 60
21. 100	22. 10	23. 40	24. 70	25. 60
26. 90	27. 20	28. 50	29. 10	30. 100
31. 30	32. 80	33. 0	34. 90	35. 0
36. 40	37. 80	38. 10	39. 50	40. 20

Answer Key # 62

1. 30	2. 60	3. 100	4. 70	5. 0
6. 60	7. 90	8. 40	9. 30	10. 50
11. 70	12. 80	13. 100	14. 10	15. 20
16. 100	17. 20	18. 90	19. 70	20. 60
21. 80	22. 50	23. 0	24. 40	25. 10
26. 30	27. 70	28. 100	29. 30	30. 10
31. 80	32. 40	33. 50	34. 90	35. 60
36. 0	37. 20	38. 0	39. 50	40. 100

Answer Key # 63

1. 80	2. 30	3. 70	4. 60	5. 90
6. 40	7. 20	8. 10	9. 20	10. 40
11. 30	12. 50	13. 100	14. 60	15. 70
16. 80	17. 0	18. 90	19. 10	20. 60
21. 40	22. 80	23. 50	24. 20	25. 0
26. 70	27. 90	28. 100	29. 10	30. 30
31. 10	32. 90	33. 80	34. 70	35. 100
36. 40	37. 20	38. 0	39. 50	40. 30

Answer Key # 64

1. 60	2. 90	3. 60	4. 30	5. 40
6. 80	7. 50	8. 70	9. 100	10. 20
11. 0	12. 10	13. 0	14. 80	15. 10
16. 100	17. 50	18. 30	19. 60	20. 70
21. 40	22. 20	23. 90	24. 70	25. 50
26. 40	27. 100	28. 60	29. 10	30. 20
31. 0	32. 30	33. 90	34. 80	35. 50
36. 20	37. 60	38. 100	39. 80	40. 0

Answer Key # 65

1. 55	2. 11	3. 77	4. 88	5. 66
6. 22	7. 0	8. 44	9. 99	10. 33
11. 110	12. 121	13. 33	14. 55	15. 0
16. 44	17. 110	18. 88	19. 22	20. 66
21. 99	22. 11	23. 121	24. 77	25. 44
26. 22	27. 110	28. 77	29. 88	30. 33
31. 11	32. 99	33. 0	34. 121	35. 66
36. 55	37. 22	38. 55	39. 44	40. 33

Answer Key # 66

1. 99	2. 66	3. 77	4. 88	5. 110
6. 121	7. 11	8. 0	9. 55	10. 121
11. 33	12. 22	13. 99	14. 77	15. 44
16. 0	17. 110	18. 66	19. 11	20. 88
21. 44	22. 55	23. 33	24. 66	25. 110
26. 99	27. 77	28. 11	29. 88	30. 0
31. 121	32. 22	33. 11	34. 0	35. 22
36. 44	37. 55	38. 33	39. 121	40. 99

Answer Key # 67

1. 88	2. 66	3. 77	4. 110	5. 66
6. 22	7. 99	8. 110	9. 0	10. 55
11. 88	12. 44	13. 33	14. 11	15. 121
16. 77	17. 33	18. 66	19. 11	20. 110
21. 121	22. 99	23. 77	24. 22	25. 88
26. 55	27. 44	28. 0	29. 11	30. 0
31. 66	32. 121	33. 44	34. 55	35. 22
36. 77	37. 99	38. 33	39. 110	40. 88

Answer Key # 68

1. 121	2. 99	3. 22	4. 66	5. 88
6. 55	7. 77	8. 11	9. 110	10. 0
11. 44	12. 33	13. 22	14. 66	15. 110
16. 44	17. 0	18. 77	19. 121	20. 99
21. 33	22. 11	23. 88	24. 55	25. 22
26. 110	27. 55	28. 44	29. 0	30. 11
31. 121	32. 99	33. 66	34. 88	35. 77
36. 33	37. 110	38. 11	39. 55	40. 88

Answer Key # 69

1. 72	2. 108	3. 48	4. 120	5. 132
6. 12	7. 144	8. 96	9. 60	10. 0
11. 84	12. 24	13. 36	14. 24	15. 108
16. 120	17. 36	18. 96	19. 144	20. 132
21. 60	22. 84	23. 72	24. 0	25. 12
26. 48	27. 96	28. 48	29. 72	30. 108
31. 132	32. 0	33. 24	34. 60	35. 36
36. 12	37. 144	38. 120	39. 84	40. 12

Answer Key # 70

1. 84	2. 24	3. 0	4. 120	5. 132
6. 72	7. 144	8. 48	9. 108	10. 36
11. 60	12. 96	13. 36	14. 24	15. 12
16. 72	17. 120	18. 84	19. 60	20. 0
21. 132	22. 96	23. 108	24. 48	25. 144
26. 48	27. 72	28. 84	29. 36	30. 132
31. 12	32. 120	33. 144	34. 24	35. 60
36. 0	37. 96	38. 108	39. 24	40. 12

Answer Key # 71

1. 120	2. 0	3. 72	4. 84	5. 96
6. 144	7. 132	8. 48	9. 60	10. 36
11. 108	12. 132	13. 108	14. 48	15. 96
16. 120	17. 24	18. 144	19. 12	20. 84
21. 0	22. 72	23. 60	24. 36	25. 60
26. 144	27. 0	28. 120	29. 36	30. 96
31. 48	32. 108	33. 132	34. 72	35. 84
36. 12	37. 24	38. 132	39. 108	40. 120

Answer Key # 72

1. 96	2. 48	3. 12	4. 144	5. 24
6. 72	7. 84	8. 36	9. 0	10. 60
11. 132	12. 24	13. 72	14. 108	15. 144
16. 84	17. 96	18. 48	19. 36	20. 12
21. 120	22. 60	23. 0	24. 36	25. 12
26. 144	27. 84	28. 132	29. 0	30. 60
31. 48	32. 24	33. 108	34. 96	35. 72
36. 120	37. 36	38. 120	39. 24	40. 0

Answer Key # 73

1. 33	2. 132	3. 84	4. 90	5. 120
6. 110	7. 48	8. 12	9. 80	10. 77
11. 100	12. 72	13. 99	14. 108	15. 0
16. 70	17. 121	18. 24	19. 30	20. 60
21. 144	22. 36	23. 20	24. 48	25. 0
26. 60	27. 132	28. 22	29. 120	30. 11
31. 55	32. 0	33. 66	34. 96	35. 110
36. 88	37. 10	38. 44	39. 40	40. 88

Answer Key # 74

1. 40	2. 12	3. 96	4. 77	5. 72
6. 132	7. 48	8. 48	9. 36	10. 84
11. 90	12. 100	13. 70	14. 0	15. 121
16. 144	17. 0	18. 110	19. 10	20. 99
21. 120	22. 55	23. 33	24. 30	25. 132
26. 44	27. 80	28. 0	29. 11	30. 60
31. 24	32. 66	33. 108	34. 20	35. 120
36. 22	37. 110	38. 60	39. 60	40. 100

Answer Key # 75

1. 33	2. 22	3. 99	4. 144	5. 48
6. 0	7. 132	8. 36	9. 120	10. 55
11. 11	12. 44	13. 40	14. 132	15. 80
16. 84	17. 96	18. 66	19. 48	20. 72
21. 88	22. 0	23. 30	24. 108	25. 120
26. 90	27. 12	28. 60	29. 121	30. 10
31. 110	32. 0	33. 24	34. 70	35. 20
36. 110	37. 77	38. 70	39. 10	40. 40

Answer Key # 76

1. 90	2. 77	3. 0	4. 60	5. 22
6. 110	7. 33	8. 80	9. 20	10. 0
11. 48	12. 121	13. 99	14. 30	15. 36
16. 60	17. 100	18. 48	19. 110	20. 24
21. 132	22. 96	23. 55	24. 0	25. 72
26. 108	27. 84	28. 144	29. 120	30. 66
31. 132	32. 44	33. 12	34. 11	35. 120
36. 88	37. 11	38. 90	39. 110	40. 12

Answer Key # 77

1. 55	2. 132	3. 22	4. 0	5. 66
6. 88	7. 48	8. 40	9. 96	10. 10
11. 20	12. 48	13. 0	14. 77	15. 36
16. 70	17. 84	18. 30	19. 121	20. 99
21. 72	22. 144	23. 60	24. 0	25. 120
26. 132	27. 33	28. 24	29. 100	30. 44
31. 80	32. 120	33. 60	34. 108	35. 110
36. 36	37. 110	38. 11	39. 60	40. 132

Answer Key # 78

1. 10	2. 60	3. 90	4. 121	5. 0
6. 20	7. 48	8. 24	9. 48	10. 110
11. 88	12. 100	13. 33	14. 44	15. 120
16. 12	17. 22	18. 144	19. 40	20. 66
21. 55	22. 96	23. 70	24. 30	25. 132
26. 120	27. 99	28. 72	29. 77	30. 108
31. 0	32. 84	33. 80	34. 0	35. 77
36. 0	37. 96	38. 20	39. 60	40. 0

Answer Key # 79

1. 90	2. 24	3. 110	4. 12	5. 44
6. 132	7. 11	8. 99	9. 33	10. 22
11. 66	12. 84	13. 50	14. 70	15. 0
16. 36	17. 50	18. 72	19. 121	20. 108
21. 88	22. 120	23. 100	24. 120	25. 30
26. 10	27. 80	28. 110	29. 60	30. 132
31. 40	32. 144	33. 55	34. 60	35. 66
36. 55	37. 60	38. 72	39. 84	40. 10

Answer Key # 80

1. 11	2. 90	3. 12	4. 24	5. 88
6. 20	7. 50	8. 99	9. 96	10. 77
11. 110	12. 100	13. 40	14. 0	15. 70
16. 33	17. 121	18. 144	19. 44	20. 0
21. 0	22. 110	23. 36	24. 132	25. 30
26. 120	27. 80	28. 108	29. 132	30. 50
31. 22	32. 120	33. 144	34. 96	35. 80
36. 36	37. 60	38. 50	39. 120	40. 132

Answer Key # 81

1. 0	2. 36	3. 0	4. 84	5. 100
6. 28	7. 88	8. 24	9. 120	10. 9
11. 120	12. 88	13. 72	14. 36	15. 48
16. 0	17. 55	18. 11	19. 0	20. 36
21. 0	22. 60	23. 8	24. 30	25. 14
26. 45	27. 56	28. 6	29. 7	30. 0
31. 30	32. 24	33. 54	34. 66	35. 70
36. 144	37. 110	38. 0	39. 12	40. 35

Answer Key # 82

1. 40	2. 63	3. 15	4. 16	5. 0
6. 121	7. 0	8. 99	9. 55	10. 16
11. 4	12. 3	13. 12	14. 36	15. 6
16. 33	17. 27	18. 64	19. 2	20. 27
21. 22	22. 21	23. 20	24. 12	25. 35
26. 96	27. 6	28. 0	29. 16	30. 45
31. 12	32. 15	33. 9	34. 28	35. 30
36. 16	37. 70	38. 6	39. 42	40. 132

Answer Key # 83

1. 24	2. 1	3. 0	4. 10	5. 48
6. 16	7. 3	8. 5	9. 72	10. 60
11. 20	12. 40	13. 12	14. 40	15. 24
16. 32	17. 48	18. 77	19. 0	20. 96
21. 0	22. 0	23. 24	24. 16	25. 0
26. 110	27. 80	28. 21	29. 44	30. 90
31. 0	32. 77	33. 90	34. 4	35. 72
36. 9	37. 36	38. 63	39. 0	40. 0

Answer Key # 84

1. 30	2. 8	3. 132	4. 33	5. 60
6. 20	7. 0	8. 108	9. 25	10. 0
11. 0	12. 32	13. 56	14. 36	15. 49
16. 42	17. 81	18. 7	19. 50	20. 72
21. 50	22. 22	23. 0	24. 18	25. 48
26. 0	27. 24	28. 0	29. 60	30. 99
31. 80	32. 2	33. 5	34. 11	35. 0
36. 8	37. 108	38. 10	39. 84	40. 0

Answer Key # 85

1. 12	2. 0	3. 44	4. 18	5. 20
6. 0	7. 10	8. 54	9. 66	10. 33
11. 30	12. 50	13. 0	14. 48	15. 24
16. 88	17. 0	18. 0	19. 60	20. 20
21. 55	22. 7	23. 8	24. 50	25. 132
26. 1	27. 56	28. 2	29. 36	30. 8
31. 84	32. 2	33. 0	34. 4	35. 18
36. 18	37. 3	38. 27	39. 10	40. 7

Answer Key # 86

1. 54	2. 24	3. 80	4. 35	5. 24
6. 10	7. 60	8. 6	9. 44	10. 24
11. 81	12. 30	13. 9	14. 6	15. 96
16. 10	17. 6	18. 4	19. 0	20. 22
21. 0	22. 12	23. 90	24. 45	25. 30
26. 110	27. 10	28. 36	29. 72	30. 72
31. 3	32. 54	33. 0	34. 4	35. 6
36. 0	37. 63	38. 36	39. 30	40. 16

Answer Key # 87

1. 0	2. 96	3. 60	4. 22	5. 21
6. 110	7. 12	8. 0	9. 32	10. 99
11. 60	12. 12	13. 15	14. 21	15. 40
16. 40	17. 120	18. 36	19. 0	20. 63
21. 121	22. 100	23. 10	24. 24	25. 12
26. 33	27. 0	28. 80	29. 0	30. 0
31. 25	32. 27	33. 99	34. 48	35. 35
36. 20	37. 40	38. 84	39. 120	40. 15

Answer Key # 88

1. 144	2. 72	3. 64	4. 66	5. 16
6. 12	7. 4	8. 24	9. 24	10. 40
11. 11	12. 5	13. 2	14. 48	15. 0
16. 9	17. 20	18. 9	19. 90	20. 0
21. 0	22. 24	23. 77	24. 66	25. 55
26. 18	27. 40	28. 4	29. 72	30. 56
31. 49	32. 0	33. 2	34. 40	35. 14
36. 77	37. 12	38. 18	39. 108	40. 40

Answer Key # 89

1. 88	2. 108	3. 14	4. 50	5. 44
6. 45	7. 28	8. 12	9. 6	10. 18
11. 132	12. 5	13. 28	14. 0	15. 70
16. 70	17. 11	18. 0	19. 28	20. 66
21. 24	22. 22	23. 25	24. 48	25. 0
26. 63	27. 42	28. 56	29. 0	30. 14
31. 2	32. 18	33. 11	34. 72	35. 56
36. 0	37. 42	38. 60	39. 0	40. 132

Answer Key # 90

1. 100	2. 110	3. 15	4. 5	5. 0
6. 24	7. 6	8. 0	9. 0	10. 36
11. 10	12. 0	13. 36	14. 24	15. 0
16. 70	17. 36	18. 45	19. 36	20. 18
21. 1	22. 0	23. 90	24. 33	25. 14
26. 60	27. 90	28. 60	29. 24	30. 55
31. 14	32. 84	33. 20	34. 7	35. 48
36. 99	37. 8	38. 72	39. 144	40. 32

Answer Key # 91

1. 16	2. 8	3. 8	4. 9	5. 0
6. 77	7. 32	8. 8	9. 20	10. 12
11. 88	12. 96	13. 8	14. 60	15. 21
16. 6	17. 12	18. 70	19. 20	20. 18
21. 99	22. 40	23. 0	24. 121	25. 49
26. 24	27. 54	28. 84	29. 2	30. 81
31. 0	32. 108	33. 16	34. 24	35. 44
36. 3	37. 44	38. 4	39. 0	40. 22

Answer Key # 92

1. 40	2. 77	3. 27	4. 0	5. 0
6. 66	7. 7	8. 14	9. 88	10. 96
11. 50	12. 9	13. 0	14. 35	15. 132
16. 10	17. 40	18. 30	19. 0	20. 35
21. 3	22. 80	23. 0	24. 6	25. 72
26. 12	27. 50	28. 0	29. 108	30. 63
31. 80	32. 28	33. 8	34. 21	35. 45
36. 18	37. 0	38. 55	39. 0	40. 30

Answer Key # 93

1. 9	2. 33	3. 40	4. 36	5. 0
6. 20	7. 15	8. 120	9. 16	10. 12
11. 54	12. 5	13. 48	14. 120	15. 110
16. 24	17. 48	18. 12	19. 72	20. 4
21. 0	22. 0	23. 12	24. 27	25. 30
26. 64	27. 11	28. 0	29. 7	30. 21
31. 3	32. 72	33. 40	34. 11	35. 48
36. 27	37. 0	38. 44	39. 15	40. 0

Answer Key # 94

1. 144	2. 5	3. 33	4. 0	5. 0
6. 88	7. 36	8. 0	9. 1	10. 12
11. 84	12. 0	13. 0	14. 108	15. 30
16. 54	17. 70	18. 110	19. 0	20. 120
21. 0	22. 54	23. 18	24. 55	25. 45
26. 6	27. 30	28. 0	29. 0	30. 84
31. 72	32. 9	33. 20	34. 60	35. 132
36. 60	37. 56	38. 99	39. 24	40. 0

Answer Key # 95

1. 0	2. 24	3. 15	4. 108	5. 0
6. 90	7. 40	8. 100	9. 0	10. 21
11. 45	12. 110	13. 99	14. 10	15. 0
16. 9	17. 24	18. 63	19. 18	20. 25
21. 0	22. 36	23. 48	24. 49	25. 33
26. 7	27. 10	28. 12	29. 64	30. 0
31. 96	32. 63	33. 48	34. 4	35. 56
36. 11	37. 28	38. 72	39. 70	40. 10

Answer Key # 96

1. 96	2. 24	3. 120	4. 42	5. 5
6. 35	7. 77	8. 81	9. 20	10. 16
11. 77	12. 30	13. 66	14. 36	15. 80
16. 0	17. 36	18. 0	19. 16	20. 6
21. 12	22. 132	23. 55	24. 80	25. 8
26. 24	27. 28	28. 2	29. 0	30. 90
31. 4	32. 36	33. 22	34. 0	35. 30
36. 22	37. 0	38. 60	39. 20	40. 10

Answer Key # 97

1. 18	2. 0	3. 0	4. 32	5. 0
6. 6	7. 60	8. 72	9. 40	10. 32
11. 18	12. 3	13. 48	14. 0	15. 48
16. 32	17. 27	18. 66	19. 35	20. 88
21. 0	22. 121	23. 18	24. 12	25. 14
26. 12	27. 20	28. 44	29. 40	30. 40
31. 12	32. 0	33. 12	34. 9	35. 48
36. 20	37. 0	38. 32	39. 99	40. 81

Answer Key # 98

1. 8	2. 0	3. 8	4. 60	5. 8
6. 90	7. 5	8. 11	9. 40	10. 8
11. 0	12. 36	13. 0	14. 72	15. 7
16. 70	17. 40	18. 36	19. 0	20. 100
21. 10	22. 88	23. 18	24. 120	25. 9
26. 8	27. 84	28. 35	29. 28	30. 22
31. 110	32. 0	33. 5	34. 0	35. 10
36. 0	37. 30	38. 132	39. 18	40. 40

Answer Key # 99

1. 108	2. 4	3. 48	4. 96	5. 16
6. 60	7. 56	8. 6	9. 0	10. 12
11. 16	12. 144	13. 96	14. 16	15. 0
16. 40	17. 36	18. 44	19. 0	20. 12
21. 27	22. 66	23. 48	24. 0	25. 120
26. 35	27. 90	28. 0	29. 12	30. 60
31. 132	32. 12	33. 0	34. 16	35. 27
36. 25	37. 99	38. 121	39. 40	40. 7

Answer Key # 100

1. 0	2. 55	3. 80	4. 30	5. 66
6. 0	7. 3	8. 44	9. 55	10. 12
11. 24	12. 9	13. 0	14. 84	15. 11
16. 108	17. 28	18. 49	19. 4	20. 0
21. 20	22. 0	23. 0	24. 36	25. 48
26. 16	27. 77	28. 63	29. 0	30. 48
31. 0	32. 4	33. 0	34. 18	35. 0
36. 63	37. 15	38. 64	39. 77	40. 72

Track your progress

Set	Time	Score	Set	Time	Score	Set	Time	Score	Set	Time	Score
1			26			51			76		
2			27			52			77		
3			28			53			78		
4			29			54			79		
5			30			55			80		
6			31			56			81		
7			32			57			82		
8			33			58			83		
9			34			59			84		
10			35			60			85		
11			36			61			86		
12			37			62			87		
13			38			63			88		
14			39			64			89		
15			40			65			90		
16			41			66			91		
17			42			67			92		
18			43			68			93		
19			44			69			94		
20			45			70			95		
21			46			71			96		
22			47			72			97		
23			48			73			98		
24			49			74			99		
25			50			75			100		

Made in United States
North Haven, CT
05 November 2026

21777131R00061